Sowing Seeds of CHANGE

Informing Public Policy in the
Economic Research Service of USDA

John F. Geweke, James T. Bonnen,
Andrew A. White, and Jeffrey J. Koshel, *Editors*

Panel to Study the Research Program of the
Economic Research Service

Committee on National Statistics

Commission on Behavioral and Social Sciences and Education

National Research Council

NATIONAL ACADEMY PRESS
Washington, D.C.

NATIONAL ACADEMY PRESS • 2101 Constitution Ave., N.W. • Washington, D.C. 20418

NOTICE: The project that is the subject of this report was approved by the Governing Board of the National Research Council, whose members are drawn from the councils of the National Academy of Sciences, the National Academy of Engineering, and the Institute of Medicine. The members of the committee responsible for the report were chosen for their special competences and with regard for appropriate balance.

Support of the work of the Committee on National Statistics is provided by a consortium of federal agencies through a grant from the National Science Foundation (Number SBR-9709489). The project that is the subject of this report is supported by Agreement No. 43-3AEJ-6-8102 between the National Academy of Sciences and the Economic Research Service of the U.S. Department of Agriculture. Any opinions, findings, conclusions, or recommendations expressed in this publication are those of the author(s) and do not necessarily reflect the view of the U.S. Department of Agriculture.

Additional copies of this report are available from National Academy Press, 2101 Constitution Avenue, N.W., Washington, D.C. 20418.
Call (800)-624-6242 or (202)-334-3313 (in the Washington metropolitan area)

This report is available on line at **http://www.nap.edu**
Printed in the United States of America

PANEL TO STUDY THE RESEARCH PROGRAM OF THE ECONOMIC RESEARCH SERVICE

JOHN F. GEWEKE (*Chair*), Department of Economics, University of Minnesota, Minneapolis

DENNIS AIGNER, Graduate School of Management, University of California, Irvine

JAMES T. BONNEN, Department of Agricultural Economics, Michigan State University

IVERY D. CLIFTON, College of Agricultural and Environmental Sciences, University of Georgia

GEORGE G. JUDGE, Graduate School, University of California, Berkeley

ROBERT C. MARSHALL, Department of Economics, Penn State University

CHARLES RIEMENSCHNEIDER, Food and Agricultural Organization, United Nations, New York

ROBERT L. THOMPSON, World Bank, Washington, D.C.

SARAHELEN THOMPSON, Department of Agricultural and Consumer Economics, University of Illinois, Urbana

ANDREW A. WHITE, *Study Director*
JEFFREY J. KOSHEL, *Study Director* (until December 1997)
KAREN HUIE, *Research Assistant*
JOSHUA S. DICK, *Senior Project Assistant*

The National Academy of Sciences is a private, nonprofit, self-perpetuating society of distinguished scholars engaged in scientific and engineering research, dedicated to the furtherance of science and technology and to their use for the general welfare. Upon the authority of the charter granted to it by the Congress in 1863, the Academy has a mandate that requires it to advise the federal government on scientific and technical matters. Dr. Bruce Alberts is president of the National Academy of Sciences.

The National Academy of Engineering was established in 1964, under the charter of the National Academy of Sciences, as a parallel organization of outstanding engineers. It is autonomous in its administration and in the selection of its members, sharing with the National Academy of Sciences the responsibility for advising the federal government. The National Academy of Engineering also sponsors engineering programs aimed at meeting national needs, encourages education and research, and recognizes the superior achievements of engineers. Dr. William A. Wulf is president of the National Academy of Engineering.

The Institute of Medicine was established in 1970 by the National Academy of Sciences to secure the services of eminent members of appropriate professions in the examination of policy matters pertaining to the health of the public. The Institute acts under the responsibility given to the National Academy of Sciences by its congressional charter to be an adviser to the federal government and, upon its own initiative, to identify issues of medical care, research, and education. Dr. Kenneth I. Shine is president of the Institute of Medicine.

The National Research Council was organized by the National Academy of Sciences in 1916 to associate the broad community of science and technology with the Academy's purposes of furthering knowledge and advising the federal government. Functioning in accordance with general policies determined by the Academy, the Council has become the principal operating agency of both the National Academy of Sciences and the National Academy of Engineering in providing services to the government, the public, and the scientific and engineering communities. The Council is administered jointly by both Academies and the Institute of Medicine. Dr. Bruce Alberts and Dr. William A. Wulf are chairman and vice chairman, respectively, of the National Research Council.

Acknowledgments

This report reflects the efforts of many people. The panel was established under the auspices of the Committee on National Statistics, directed by Miron Straf, who was instrumental in developing the study and provided guidance and support to the staff.

The panel gratefully acknowledges the funding received from the Economic Research Service (ERS) of the U.S. Department of Agriculture. The administrator of the ERS, Susan Offutt, and its division directors and other senior staff provided publications, documents, information, and consultation that were essential and helpful in the panel's work.

The panel benefited greatly from consultation with current and former agency heads, economists, other social scientists, and historians who gave generously of their time. In particular we would like to thank Harry Ayer, Sandra Batie, J. Roy Black, William Browne, David Chu, Willard Cochrane, Lynn Daft, Kenneth Deavers, Clark Edwards, Brian Fisher, Dale Hathaway, James Hosek, John Lee, Jr., John Miranowski, Janet Norwood, Wayne Rasmussen, Sherman Robinson, Vernon Ruttan, John Schnittker, David Schweikhardt, Matthew Shapiro, Ed Simms, David Smallwood, Joshua Sosland, David Sundig, Laurian Unnevehr, Abner Womack, and Gene Wunderlich, as well as others who wished to remain anonymous.

The panel's task could not be completed without an excellent, well-managed staff. In particular, the report would not have been possible without the efforts of a number of staff members. The panel is indebted to Joshua Dick, senior project assistant, who skillfully guided the report through numerous rounds of editing in preparation for publication and ably and cheerfully handled the administrative tasks presented by the panel. Karen Huie, research assistant, provided valuable research assistance. Patricia Eisele and Beryl Watters at Michigan State University patiently keyed and rekeyed numerous versions of the history of ERS. Finally, we are indebted to our study directors. Jeffrey Koshel managed the overall work of the panel and provided counsel during the first half of the panel's work. Upon his departure from the Committee on National Statistics, Andy White ably stepped in to manage activities and help us to meet our deadlines. Together Jeff and Andy worked on all aspects of the project and were always dedicated, responsible, and in good humor.

This report has been reviewed in draft form by individuals chosen for their diverse perspectives and technical expertise, in accordance with procedures approved by the Report Review Committee of the National Research Council. The purpose of this independent review is to provide candid and critical comments that will assist the institution in making the published report as sound as possible and to ensure that the report meets institutional standards for objectivity, evidence, and responsiveness to the study charge. The review comments and draft manuscript remain confidential to protect the integrity of the deliberative process.

We wish to thank the following individuals for their participation in the review of this report: David S.C. Chu, RAND, Washington, D.C.; Kenneth R. Farrell, Vice President for Agriculture and Natural Resources, University of California, Berkeley; Bruce L. Gardner, Department of Agricultural and Resource Economics, University of Maryland; Eric Hanushek, Wallis Institute of Political Economy, University of Rochester; Janet Norwood, Urban Institute, Washington, D.C.; Vernon W. Ruttan, Department of Applied Economics, University of Minnesota; Francisco J. Samaniego, Department of Statistics, University of California, Davis; and G. Edward Schuh, Hubert H. Humphrey Institute of Public Affairs, University of Minnesota.

Although the individuals listed above have provided constructive comments and suggestions, it must be emphasized that responsibility for the final content of this report rests entirely with the authoring committee and the institution.

I close by expressing my appreciation to fellow panel members for their willingness to devote long hours to this project. Coming from diverse backgrounds, they worked together well and patiently. A number of panel members prepared drafts for the panel's use. James Bonnen, in particular, devoted many weeks to drawing together hundreds of sources to provide a comprehensive background and history of ERS and its predecessor agency, the Bureau of Agricultural Economics. All panel members responded in an extremely helpful way to the requests I made of their time.

John Geweke, *Chair*
Panel to Study the Research Program
of the Economic Research Service

Contents

Sowing Seeds of Change

Executive Summary

Economic policy making is an inescapable activity of government in a representative democracy with an economy grounded in free market principles. In a world of constantly evolving technology and information, most social institutions, including markets and property rights, are also changing. Many markets function well, but only government can provide the legal framework in which these markets exist. Changing technology leads to new markets, for which changes in this framework are required. Some markets do not function well because of the inherent characteristics of the goods and services traded in these markets. For these markets, too much or too little of the commodity will be produced or consumed unless government intervenes. In this environment, new research and information must constantly be brought to bear if economic policy is to be made wisely. The quality of economic policy decisions affects the welfare of the nation's individuals, and is an important factor in the competitive position of our nation with respect to others.

Government agencies charged with policy support responsibilities are some of the most important conduits from new research and information to public economic policy. Within the U.S. Department of Agriculture (USDA), the Economic Research Service (ERS) as did its predecessor, the Bureau of Agricultural Economics (BAE), from 1922 to 1953, provides much of the research and information in support of the department's economic policy mandate. In early 1997, ERS requested that the Committee on National Statistics of the National Research Council convene a panel to assess the management and structure of the ERS research program and produce a report of general principles for improving the quality and effectiveness of research in an intramural social science program that must serve agency program needs. The panel was also asked to examine

1

attendant issues, including relevance, timeliness, quality standards, and employing research by others, and to recommend changes in the management and structure of the research program.

In meeting its charge, the panel undertook several initiatives to understand the function of ERS and similar agencies in the policy-making environment; to evaluate the research and information services of ERS with respect to quality, relevance, timeliness, and credibility; to study alternative organizations for providing research and information in the United States and other developed countries; and to evaluate the management of ERS. The panel studied the history of ERS since the establishment of the BAE in 1922, examined the economics of the supply of research, information, and analysis, and considered the scope for potential change in ERS and similar agencies. The panel carried on extensive discussions with current and former administrators of agencies; conferred with current administrators and senior staff at ERS; consulted with clients of ERS; read ERS research reports, staff analyses, and publications; reviewed the relevant theoretical and practical literature; and carried on extended internal discussions. The panel met five times between June 1997 and June 1998. This report reflects the program and organization of ERS as of early 1998.

This report presents the panel's findings, conclusions, and recommendations. Consistent with the charge to the panel, the recommendations address ERS, the Office of the Secretary of Agriculture, and the committees of Congress that set policy and appropriate funds for ERS. They also bear on other agencies providing the information and organizing the research and analysis that is necessary to inform public policy. The report sets forth a goal for informed public economic policy that the panel believes is attainable and recommends essential changes to realize this goal. This summary contains the most urgent recommendations; the complete set of recommendations is found in the body of the report.

THE ERS AND INFORMED PUBLIC ECONOMIC POLICY

Since 1922 the BEA and ERS have provided research and information to support the policy mandate of USDA, which has evolved extensively in the intervening 77 years. The most immediate and visible category of service is staff analysis in response to questions from the Office of the Secretary, and often from the Office of the Chief Economist. Other requests come from USDA agencies, the Congress, the Office of Management and Budget, and the Council of Economic Advisers. In 1997, ERS utilized about 20 percent of the time of its staff of professional employees, most of whom are economists, responding to about 350 such requests.

The second category of service is the development of secondary data and analysis, often presented in the form of indicators and accounts, such as those found in USDA *Situation and Outlook* reports. This activity occupies about 40 percent of professional staff time. Indicators and accounts provide quantitative summaries of economic activity in the food, fiber, and natural resource sectors of the economy, which account for about 16 percent of gross domestic product. Examples include agricultural trade trends and forecasts, farm financial status, food consumption and waste, marketing margins, and natural resource use in agriculture and associated indicators of environmental quality.

The final category of service is intermediate and long-term research related to the economic policy mandate of USDA. This research reflects the diversity of the economic policy mandate of USDA, including, for example, economic incentives for potential participants in the Conservation Reserve Program, evaluation of commodity procurement for food assistance programs, analysis of the economic impacts of proposed changes in tariffs on agricultural products, and cost-benefit analysis of conservation tillage.

In the past 20 years, the breadth of research, information, and analysis required in ERS has grown along with the policy mandate of USDA. Environmental and food safety issues, global warming, the consequences of international financial instability, and other issues have augmented the traditional concerns with production agriculture. Yet inflation-adjusted funding has decreased by one-sixth and staff by over 30 percent since 1992. The training of ERS professional staff reflects its traditional agenda more than it does the current policy mandate of USDA. Whereas once ERS was the dominant employer of agricultural economists with limited alternative career prospects, today ERS must compete to recruit and retain a much broader array of professionals, most of whom have many attractive alternatives. All of these factors, combined in a fractious political environment, have led to a widespread perception that the quantity and quality of ERS products are not what they should be, the reduced staff and budget of ERS perhaps notwithstanding. These reservations about ERS are becoming more acute, as ERS is asked to address an ever-widening range of serious economic issues. The recommendations in this report address these problems specifically, as well as general principles for improving the quality and effectiveness of agencies charged with producing research, information, and analysis in support of economic policy.

EVALUATION AS A FRAMEWORK FOR MANAGEMENT

The principle of competitive supply applies to the provision of research and information in support of public economic policy, just as it does to other economic activities. The ultimate consumers of these services compare alternative sources and make choices. A successful provider, including ERS, must clearly identify the services it provides and the clients for its services. The successful provider must also understand who its competitors are or could be, and the attributes of the service that underlie the comparisons and choices of its clients. Like any successful enterprise in a competitive market, an agency providing research and information in support of public policy must continually evaluate itself retrospectively on these four dimensions: the services it provides, the set of current and potential providers of these services, existing and potential clients, and the attributes of services that underlie clients' comparisons and choices. Employing the evaluation process prospectively, such an agency should manage itself so as to achieve the most favorable ultimate evaluation.

On the basis of its examination, the panel concludes that there are four attributes of research, information, and analysis that matter to public policy makers and other public- and private-sector clients. First, research, information, and analyses should be of high quality, meeting relevant disciplinary and professional standards. Second, research and analyses should be relevant, addressing the essential policy question and with consideration of the policy context in which decisions are made. Third, these services should be timely. Intermediate and long-term research conducted in anticipation of policy questions and concluded before political lines are drawn is a treasured resource, not only for clearly being independent of specific interests, but also for its availability at critical junctures, when decisions must be based on what is known rather than what might be learned. Fourth, all of the services provided by agencies in support of public economic policy must be credible. The credibility of research, information, and analysis in support of public economic policy derives from its quality, relevance, and timeliness, and its established independence from the political decision-making process.

Until recently, agencies that support public economic policy have not been evaluated formally with respect to these dimensions. The panel finds that there is little infrastructure for effective evaluation in ERS, including its responses to the requirements of the Government Performance and Results Act of 1996. For example, ERS administrators are unable to account for staff time and other resources on a project-by-project basis, even on a large scale.

The panel recommends that ERS systematically evaluate the services it provides. Formal program evaluation instruments should elicit from clients and potential clients their choices among alternative providers and potential providers of the services provided by ERS, and the attributes of the services critical to their choices, including prices. The instrument should solicit the identities of additional potential clients and alternative providers of these services. ERS should participate in the design of evaluation instruments, but their administration should be delegated to an independent party. The panel further recommends that ERS should allocate its costs and staff time across the same services used in its system of evaluation, according to generally accepted accounting principles.

In the long term, ERS or any agency providing research and information in support of public economic policy must have the widest possible scope for the way it produces these services. In its review of these agencies and discussions with their current and former administrators, the panel found a rich array of organizations, including long-term relationships with university-based research institutes and federally funded research and development centers; the use of grants, contracts, and cooperative agreements with individual and small groups of investigators; quasi-public agencies; mixtures of intramural and extramural research; and various forms of public-private partnerships; as well as intramural research and information programs carried out exclusively by permanent agency employees. These organizations reflect different methods for procuring the research and information that supports public policy. A successful agency must be able to choose among these methods of procurement over the long term, and it must be free to make changes as its policy mandate and operating environment evolve. Specific choices should be reflected in agencies' strategic plans, but not their more durable mission statements.

The panel recommends that the mission of ERS should be to provide timely, relevant, and credible information and research of high quality to inform economic policy decision making in USDA, the executive and legislative branches of the federal government, and the private and public sectors generally. It should identify information and frame research questions that will enhance and improve economic policy decisions within the authority of the secretary of agriculture, organize the subsequent collection of information and conduct of research, and evaluate alternative approaches to policy problems. The work of ERS should address anticipated as well as current and continuing policy questions.

ADMINISTRATION OF RESEARCH, INFORMATION, AND ANALYSIS

The principle that services should be supplied competitively is a fundamental premise of our economic system, including government procurement. This principle is recognized in the Competition in Contracting Act of 1984. A competitive supplier of research services, in particular, must constantly integrate new ideas in order to continue supplying those services. There is pressure for both individuals and organizations to reach beyond their immediate area of expertise to gain a competitive edge by bringing to bear new results from related fields. Suppliers of research services with permanent, sole-source awards have no such incentives to reach out and often become isolated within narrow fields using methods that are increasingly outdated.

The panel concludes that no organization should ever be given a permanent sole-source award for the provision of a service. Decisions to provide sole-source awards must be defended on a recurring basis, beginning from the presumption that services should be procured competitively.

The panel recommends that research and information in support of public economic policy should be procured competitively. All potential suppliers, including ERS, should be on the same competitive footing. If an outside supplier is selected as an awardee, in many cases ERS should have a secondary role as a partner in the provision of the service. No supplier, including ERS, should have a permanent, sole-source award for the provision of any service. Any decision to grant a sole-source award must be defended periodically.

The partnership role of ERS envisioned in this recommendation is important to the effective competitive procurement of research and information in support of public economic policy. It is essential that USDA ensure that the supplier has an understanding of the policy context for research and information and apprise contractors of developments in policy over the lifetime of the award. Intellectual command of the areas in which research is contracted, as well as an understanding of the universe of potential suppliers and their capabilities, are essential to making appropriate decisions about the procurement of research and information. In many cases, a partnership role for ERS may be the most effective way for USDA to ensure that it effectively addresses the critical choice of the best vendor for the research and information needed to support its policy decisions.

The panel recommends that choices among alternative vendors of research and information in support of public economic policy should be based on prospects for favorable evaluation of the services provided, as well as on the costs of the services. The critical attributes established in program evaluation provide the framework for choice among vendors. No single model of choice among vendors is appropriate for all programs. In particular, the methods used by the National Science Foundation, the National Institutes of Health, the Agricultural Research Service, and the National Research Initiative will not be suited to many ERS programs and should not be presumed to be appropriate to any.

Expanding the universe of potential suppliers of research and information is essential to the success of ERS given the breadth, growth, and changing character of the USDA economic policy mandate. It is unlikely that any permanent staff of professionals could, under foreseeable budgetary conditions, come close to meeting the needs for information and research in support of economic policy in USDA. The flexibility in choosing suppliers demanded by the principle of competitive procurement will also enable USDA to meet expectations of quality and quantity in the research, information, and analysis provided by ERS.

Intermediate and Long-Term Research

The credibility of its research has been troublesome throughout the history of the ERS and its predecessor agency, the BAE, as well as in other agencies providing research and information in support of public policy. Separating findings of fact from political considerations in a credible fashion is one of the most difficult tasks in the administration of research and information production. Identifying and studying relevant policy problems before they emerge as political issues enhances credibility, but achieving such success on a regular basis is not realistic given practical limitations on resources.

The most important practical consideration is to distance those who provide counsel to decision makers, like the secretary of agriculture, from those who carry out the research in support of their decisions. This fact has been demonstrated repeatedly in the history of the BAE and ERS, and it has been identified previously by others who have analyzed the problem.

The separation of the conduct of research from the making of policy has both administrative and organizational implications. With respect to administration, the most difficult problem is the need for research and information findings to be cleared by political appointees in government agencies. The panel finds that there is no effective substitute for the independence of research from the making of policy.

The panel recommends that USDA should support the integrity of its intermediate and long-term research programs in support of economic policy, while retaining the prerogative to disagree with research findings. These programs should be conducted with the clear objective that peer-reviewed research findings may be published by the investigators independently and without prior approval by USDA, and with the clear understanding that USDA does not necessarily endorse the findings of any research program.

If the independence of intermediate and long-term research conducted by ERS employees from the political process cannot be guaranteed in this way, then this research should be carried out by external vendors to whom these guarantees can be extended, as they are, for example, when agency-sponsored research is published in an academic journal with agency disclaimer. Peer reviews of research are always appropriate. The model of peer review, not prior clearance, will be more effective in obtaining the service of the best professionals (whether employees or external vendors), a step that is essential to a reputation for quality. The maintenance of the integrity of government-sponsored research is the responsibility of the research agency, its cabinet secretary, and the relevant committees of Congress.

Reports and Indicators

Both historically and currently, there has been consistent demand for the provision of secondary data with accompanying analyses by ERS in the public and private sector, and there has been more consistent political support for this function of ERS than for its intermediate and long-term research. These services are vital to USDA in providing detailed projections of supply, demand, and prices in order to estimate the budgetary and farm income implications of extensive price support programs. But these needs are changing. Many of these programs are being reduced under farm legislation enacted in 1996 and may or may not be further reduced or even eliminated in 2002, the next farm legislation renewal date. At the same time, environmental and other regulations in agriculture are increasing, providing a renewed justification for some existing indicators and secondary data as well as creating new needs for USDA analysis.

The provision of price and production reports and indicators by USDA extends well back into the nineteenth century, predating commodity support programs, which began in the 1930s. The modern program of market and farm income outlook reports began in the 1920s. USDA reports and analysis were developed to provide farmers with market information similar to that available to purchasers of their commodities. Political support for the provision of secondary data, information, and accompanying analyses by ERS has been grounded in these considerations of equity. In the wake of the information revolution, valid arguments for the public collection and provision of primary data—including increasing returns to scale and the pure public good nature of these data—may or may not apply in the same way to secondary data and information. New public policy demands in areas like nutrition and food safety may also change government needs regarding indicators and secondary data. In light of all of these changes, the reports and indicators programs must be reexamined from first principles.

The panel recommends that the secondary data preparation and analysis programs of ERS should be evaluated within the framework outlined by the panel, including consultations with clients. On the basis of this evaluation, a long-term plan should be drawn up, including new and discontinued services. The plan should indicate which of the services provided will be produced in ERS, which will be procured from other vendors, and which will be left to the private sector. The plan should include anticipated impacts on clients and the projected impact on the USDA budget.

Staff Analysis

Staff analysis is the point of contact between the ERS research, information, and analysis programs and the policy decisions that these programs support. Close contact between staff analysis leaders and policy makers is required to ensure that the entire ERS program remains relevant to the substantive economic policy mandate of USDA. Information that must be provided on a very short-term basis—often a few days or less—requires that those providing the information be available immediately. Highly political requests should go to the Office of the Secretary or the Office of the Chief Economist.

Staff analysts must also be closely involved in guiding the ERS research and information program, including the assessment of future policy questions, the framing of questions for investigation, and the organization and supervision of research, because staff analysts are the first line of contact with policy decisions. Thus, leadership in staff analysis requires a sophisticated combination of analytical and management skills. The important attributes of credibility and relevance in staff analysis and the need for leadership in staff analysis to oversee research and information programs indicate that this function must be provided by a permanent, skilled group of staff analysts within USDA.

The panel recommends that USDA should maintain a permanent core of staff analysts to provide immediate support for its economic policy decisions. The size and composition of this group should reflect the level of detail and timeliness required in support of the economic policy mandate of USDA, and it should be reviewed from time to time as the mandate evolves. The leadership of this group must provide a combination of management and analytical skills essential to the administration of the research and information programs of ERS. ERS should regularly invigorate this group by means of visiting scholars, sabbaticals, internships, or similar programs, to maintain the contact of staff analysts with the wider research community.

ORGANIZATION AND PLACEMENT

The mission of ERS and evaluation of the services it delivers drive the administration of research, information, and policy that we recommend. This administration of services will be effective in delivering research and information to policy makers only if it is embedded in organizations that support it, extending from USDA to the president and the Congress. As the history of this function in the USDA shows (see Chapters 3 and 4), this has not always been the case.

ERS within USDA Today

In the current organization of USDA, the Office of Chief Economist, situated within the Office of the Secretary, has primary responsibility for economic policy advice to the secretary. The chief economist is appointed to serve the secretary, has direct contact with the secretary in policy meetings, and has a small policy advisory staff of about eight professionals. The administrator of ERS, in contrast, reports to the Under Secretary for Research, Education, and Economics (REE), along with the administrators of three much larger agencies—the Agricultural Research Service, which oversees largely biological research, the National Agricultural Statistical Service (NASS), which collects primary data, and the Cooperative State Research, Education, and Extension Service. NASS collects much of the primary data used by ERS, and ERS is one of the principal clients of NASS. It is a statistical agency, and most of the data it collects are economic. The REE undersecretary has no responsibility for economic policy and is not likely to be a social scientist. Many requests to ERS for staff analysis come from the chief economist.

Research and information in support of economic policy within USDA are not well served by these lines of authority. The administrator of ERS, with responsibility for over 300 professional employees, is several steps removed from the policy process to which the work of ERS must be relevant. The chief economist, charged with representing economic information in the decision-making process, has no direct line of authority to the greatest concentration of talent in USDA for marshaling this information.

These lines of authority would not serve research and information in support of economic policy well under the model of competitive procurement of services by ERS advanced in this report, either. In the current organization, there is no position suited to deciding whether particular information and research services in support of economic policy should be procured from outside vendors, or, in the event that both ERS and outside vendors could supply services, whether or not ERS should be chosen. Reorganization of the economic policy support function within USDA should therefore be considered simultaneously with the question of how these research and information services are procured.

Recommended Reorganization

The principles for procuring information and research, the history of the BAE and ERS, and the experience of other cabinet-level agencies suggest a reorganization that copes with all of these problems. First, both economic policy decision making and research and information in support of economic policy should be brought into a single line of authority. This was the case for many years in the BAE and ERS, and it is true in many cabinet-level departments today. Second, consistent with the lessons learned from the history of the BAE and ERS and with the model for procurement of information and research services developed in this report, the functional separation between policy decisions, on one hand, and credible research of high quality in support of these decisions, on the other, should be clear and transparent.

The panel recommends that ERS should never be involved in recommending or deciding on specific policy actions, which are the prerogative of the secretary.

The panel recommends that a small, highly capable policy analysis and advisory group should be led by an appointee, such as a chief economist or an assistant secretary for economics, who manages day-to-day economic policy staff support for the Office of the Secretary. Such a unit would be appointed to serve the secretary and would provide any advice on political and policy action, keeping prescriptive advice and highly political matters from being directed to ERS. The administrators of the Economic Research Service and the National Agricultural Statistical Service should report to the chief economist or the assistant secretary for economics.

The Office of the Chief Economist or the Assistant Secretary for Economics, in this recommended organization, requires individuals with a thorough understanding of current and emerging policy issues and strong abilities in framing research questions. The Office of the Chief Economist or Assistant Secretary for Economics must be able to pose well-framed research requests that address their policy needs, while balancing timeliness, qualifications to do the work, and a sense of what is possible.

The professional staff in the Office of the Chief Economist or the Assistant Secretary for Economics, and not ERS, would be responsible for bringing the research and information services of ERS to bear in policy councils. They must therefore have a thorough command of the economics of policy questions, whether provided internally by ERS, through sponsored extramural research, or through syntheses of existing research and information. The same staff of the Office of the Chief Economist or the Assistant Secretary for Economics would be responsible for evaluating the program of research and information conducted externally and through ERS, for directing research and information projects, and for choosing vendors for research and information.

ERS, in this recommended organization, would have primary responsibility for the policy relevance of research programs in its role as primary or secondary provider, would be responsible for the administration of internal research and information projects, and would have a direct interest in maintaining programs that are competitive with alternatives in the public, private, and academic sectors. The administrator of ERS should be a professional, career economist, not subject to political appointment. He or she would be available to explain the research and information findings of ERS, as would external contractors, but should never be called on to represent the policy position of the secretary, the assistant secretary for economics, or the chief economist.

CONCLUSION

Adoption of the recommendations in this report will be effective only if there is agreement among senior policy makers on the principal points underlying them. These points include the nature of public economic policy and the desirability of informed rather than uninformed policy. In the production of information, research, and analysis to inform public economic policy, they include the principle of competition and the necessary attributes of quality, relevance, timeliness, and credibility.

The history of ERS amply demonstrates the vulnerability of an agency that informs policy decisions with credible and relevant information yet is not itself a political decision maker. Yet the same history indicates that this role is essential to success in informing policy decisions. The concept of such an agency is too fragile to sustain disparate expectations by the executive and legislative branches. It requires cooperation and agreement between the secretary and the relevant congressional leadership on a common set of expectations and rules for shared access to ERS services and the role and expected behavior of ERS in dealing with both branches of government. Only in such an environment will informed public economic policy survive.

1

Overview

Economic policy making is an inescapable activity of government. In a world of constantly evolving technology and information, many social institutions, including markets and property rights, are also changing. Many markets function well, but only government can provide the legal framework in which these markets exist, and changing technology leads to new markets, for which changes in this framework are required. Some markets do not function well because of the inherent characteristics of the goods and services traded; for these markets, too much or too little of the commodity will be produced or consumed unless government intervenes. In this environment, new research and information must constantly be brought to bear if economic policy is to be made wisely. The quality of economic policy decisions affects the nation's well-being and is an important factor in its competitive position with respect to other nations.

New research and information are produced at many points in the public and private sectors, and public economic policy is now made at international, national, state, and local levels. Many markets and firms are global. The interaction of research and policy occurs in diverse ways. One of the most important is the intramural programs of economic research and information in government agencies charged with responsibilities for support of economic policy. Within the U.S. Department of Agriculture (USDA), the Economic Research Service (ERS), and before 1953 its predecessor, the Bureau of Agricultural Economics (BAE), has conducted such a program.

At the request of ERS, the Committee on National Statistics of the National Research Council convened a panel to study general principles for improving the quality and effectiveness of research in an intramural social science program that must serve agency program needs. The panel was asked to examine a variety of subsidiary issues, including quality standards, evaluation, resource allocation, and employing research performed by others. ERS requested that the panel recommend changes in the management and structure of the research program in order to improve research quality. This report presents the panel's findings, conclusions, and recommendations.

The report begins by laying out the issues that underlie public economic policy. Chapter 2 raises several questions: What are the principal reasons that governments should or do intervene in market economies in representative democracies? Why not simply apply laissez-faire principles universally? The economic characteristics of markets that lead to bad outcomes under laissez-faire principles are well understood, and this understanding successfully predicts the outcomes of different government interventions. Attempting to intervene without understanding these characteristics is likely to lead to bad outcomes. Understanding the relevant economic characteristics in any particular market requires research planners to develop a useful analytical framework. In addition, good data and other information are needed to design and implement specific policies. The demand to understand the policy implications of changes in markets that are driven by innovations in science and technology presents a need for the continued development of new information and research in support of economic policy.

Producing data, information, and research in support of policy is itself an economic problem. The same economic principles that predict the outcomes of government interventions in markets also indicate when private markets will produce these needed data, information, and research. In most cases they will not, or they will produce too little, or they will produce data, information, and research that are not useful in public policy. Public sponsorship is therefore needed. Economic analysis can be used to trace the implications of different kinds of public sponsorship, through the incentives created for the individuals and organizations involved in producing information and research.

The development of agriculture in the United States provides an enlightening case study of government intervention in changing markets. It is rich with examples that inform our understanding of this process. Changes in agricultural technology, founded in both the physical and life sciences, have increased both production per worker and production per acre more than tenfold in the past century. Technological changes have differed in form and degree over the hundreds of agricultural commodities, each with its own market. Since a significant portion of agricultural biological technology is specific to location, changes have differed by substate regions of the United States as well. There has been a long-standing commitment to public-sector research and information production in U.S. agriculture, not only in the physical and life sciences, but also in economics, to support the understanding of public policy appropriate to agriculture as markets have changed with technology. In economics and statistics, this commitment can be traced at the federal level to 1840. Growing research and information production in economics led to the creation of the Bureau of Agricultural

Economics in the U.S. Department of Agriculture in 1922. It was the predecessor agency of the ERS.

The collective history of the BAE and ERS, described in Chapter 3, is rich with different structures for managing the production of economic information and research in support of private decisions and public policy in the agricultural sector. Some of these experiences were successful, others less so, and some of the same arrangements have been tried more than once in the past 75 years. From these experiences, a number of lessons can be drawn. The chapter concludes with these, and Chapter 4 goes on to lay out some of the problems currently facing ERS and USDA.

Research and information agencies in support of public economic policy, of which the ERS is one, supply a variety of services to the departments in which they are located, as well as to other agencies, the Congress, and the public. They provide analytical support for decision making, tracing through the likely implications of potential changes in policy. They often collect primary data, or transform primary data and combine it with information in a way that strengthens their decision-making support function. They often conduct or sponsor intermediate and long-term research that will provide the foundation for improved analysis and data collection at the time it is needed to support policy decision making. For these various services, the agencies involved have different clients in government and the private sector, and they have numerous options for procuring the data, information, and research. Some agencies produce information and research internally, others rely on grants and contracts, and yet others develop long-term relations with organizations outside government that do much of the actual work.

To be effective, a research and information agency in support of public economic policy must understand the nature of the services it is charged to provide, the clients to whom it provides the services, and the way that its clients will evaluate the services provided. Recent changes in management policy within the federal government, including the Government Performance and Results Act of 1996, require that these understandings be made more explicit than they have been in the past. To the extent that the management of an agency clearly perceives its services, its clients, and the attributes on which its delivery of services will be evaluated by its clients, it can prospectively consider different arrangements for delivering services and for managing the delivery of its services.

This report provides specific details on how ERS—and by extension, any research and information agency in support of public economic policy—can implement this process. It is fundamental that the choice among different vendors for services—for example, permanent branches in the agency, independent institutes under long-term contract to the agency, and individual grantees or contractors—be made on the basis of the attributes of the services that will ultimately be evaluated. Chapter 5 explains this prospective use of the evaluation process and extends it to evaluating the performance of staff within the agency.

In applying these principles to ERS, the report draws on the rich history of the agency and its predecessor, the BAE, on the mix of services currently provided by ERS and the way these services are changing, and on the various ways that research and information in support of public economic policy are produced and organized elsewhere in the federal government, the United States, and the world. Chapter 6 identifies four attributes of leading importance to most of the clients for most ERS services: timeliness, relevance, quality, and credibility. The report then poses the question, how should the production of information and research be administered by ERS management, so that it compares favorably with alternative arrangements, in all four attributes? Among a great number of possibilities, the potential answers include: do more of the same, do it differently, or don't do it at all. The outcomes are far from the same for different ERS services; for example, the implications for day-to-day staff analysis of current policy questions and for intermediate and long-term research, are quite different. In each case, however, the principle that potential vendors for all services must be competitive is maintained: no organization is ever given a permanent retainer for the provision of a service—whether it is a branch within an agency or an institution outside an agency.

The organizational framework within which a research and information agency in support of public economic policy is placed vitally affects its ability to marshal and administer its services effectively. This framework must cope with a tension that is evident in all government agencies charged with economic policy responsibilities and throughout the history of ERS. On one hand, the research and information produced or procured by such an agency must be relevant to the policy questions of the department in question. To this end, close contact with policy makers is important. On the other hand, political involvement of researchers, and those administering research, is invariably damaging to the quality and credibility of the work in the long run (and often in the short run). The final chapter of the report applies the principles of evaluation and administration developed earlier to this question, making recommendations about the organization within USDA and for determining the size and scope of the responsibilities of ERS.

Throughout the report we address arrangements for producing research and information that will best serve public economic policy in general and federal economic policy for food, agriculture, and natural resources in particular. Pursuing the recommendations made here will require effort, perseverance, and time. The report provides a goal of achieving the best information and research support possible for economic policy within resource constraints. It provides a broad framework for achieving that goal. How that goal is pursued, and how long it will take to achieve, will depend in no small part on current political and economic considerations. Along the way, some of the goals described here will not be met. It is the hope of the panel that identifying the goal will facilitate movement in its direction.

2

Informed Public Economic Policy

Public policy making in the United States is a large and complex undertaking. In 1997, new legislation amounted to some 2,691 pages and changes in regulations required 68,530 pages of the *Federal Register*. Much of this activity is devoted to economic concerns, and the laws and regulations are means to many different economic and social objectives. The link between policies and economic objectives is often difficult to anticipate. For this reason, policy makers often seek to inform their understanding of the effects of policy changes before changes are put into effect.

As the issues confronting society and government have become more complex, so too has the study of effective economic policy. In the last 50 years, the increased involvement of government in economic policy has been accompanied by more detailed collection of data, expanded production of information of many kinds, and intensified economic analysis of the effects of alternative policies. In fiscal 1996, some 29 agencies of the federal government carried out significant economic information collection or research in support of their mission, at a cost of $182 million.[1] The Economic Research Service (ERS) of the U.S. Department of Agriculture (USDA), in its previous manifestation as the Bureau of Agricultural Economics, is one of the oldest agencies of the federal government whose primary mission is providing economic research and information in support of

---

[1]National Science Foundation spending of about $18 million in economics is excluded from this figure. An additional $432 million in social science research is not explicitly classified, so total spending for economics is in all likelihood actually substantially higher (National Science Foundation/SRS).

both public- and private-sector decision making. In fiscal 1996 its budget was nearly $53 million.

To understand and evaluate the function of any such federal agency, it is necessary to first appreciate the reasons for public economic policy in a representative democracy with a market economy. These reasons drive the policy-making agenda. The provision of research and information in support of public economic policy is itself an economic activity. Whether the right amount of research and information will be provided by the private sector, and, if not, how government should intervene, is itself a very interesting question of public economic policy. The answer depends on the relevant characteristics of this economic activity. Much of this report is an analysis of these characteristics and their implications for how the production of research and information in support of public economic policy should be organized. The themes set out in this chapter reappear subsequently in this report in the detailed consideration of the specifics of the ERS.

THE NATURE OF PUBLIC ECONOMIC POLICY

Governments regularly intervene in market economies in representative democracies when the conditions that are necessary for markets to produce an efficient and equitable distribution of resources do not exist. Although specific reasons for intervention are many and vary with times and issues, many interventions can be ascribed to one of several kinds of actual or perceived failures of markets to produce efficient outcomes. Sources of these failures include natural monopolies, externalities, public goods, barriers to information, ill-defined property rights, and considerations of equity.

Natural Monopolies

Markets can lead to inefficient production levels in a particular industry, if it is technically most efficient for the good produced in that industry to be provided by a single firm. This will happen if the cost per unit of production continues to go down as more of the good is produced. For example, it was technically inefficient for more than one railroad to provide service to a local community, in most cases. The market outcome is a monopoly firm, which then charges a higher price than would be charged in a competitive industry. Consumers will demand less, and too little of the good will be produced. The usual solution is that the industry is regulated, or even owned, by the government, with the objective of producing a more efficient volume of the good. In the case of nineteenth-century railroads, the Interstate Commerce Commission was created to oversee freight rates. The cost of information sometimes continues to go down as more of it is produced by the same organization. For example, the Census Bureau produces the large national censuses and several household surveys. These products are widely used in the private sector as well as for public policy making.

Externalities

Markets can lead to inefficient production levels because the decisions of an individual or an organization cannot be excluded from affecting the economic interests of other individuals or organizations. This kind of failure is called an externality. For example, if a large hog-raising enterprise is set up in a rural community, there may be consequent air and water pollution that adversely affects nearby residents. Various farming activities may degrade water quality if carried out in sensitive watersheds. In such cases, government intervention may be able to transfer costs borne by others to decision makers, thereby leading to a more efficient resource allocation. This may take the form of defining property rights: for example, if it is established that the rural community owns the water rights, then the community and the hog-raising enterprise may come to terms about water quality and compensation.

Public Goods

If a commodity can be consumed by many individuals, and consumption by one does not reduce the opportunity for consumption by others, then the commodity in question is said to be nonrival and is a public good. An important class of examples is new information that contributes to the productivity in an industry, with many competitors, like farming. If it is not possible to charge each individual for his or her consumption of the commodity, the commodity is nonexcludable. Commodities that are both nonrival and nonexcludable are pure public goods. Although pure public goods are demanded by many individuals, it is difficult to get individuals to pay for them voluntarily. If others pay, then a given individual who does not contribute toward provision of the good is still able to consume the good in exactly the same way and extent as if they had contributed. National defense, and in many instances important new information, are leading examples of pure public goods. Too little of a pure public good will be produced because it is difficult to get individuals or firms to pay. Government may intervene by producing the good directly, for example by providing education about new farming practices or statistics on prices and production. It might intervene indirectly by redefining property rights to embody the information in an excludable good, for example by creation of a patent system. Since an informed public is essential to the performance of democratic institutions, and since many forms of information are pure public goods, the potential for underproduction of information in a market-oriented democracy is a public policy issue of the highest magnitude.

Barriers to Information

Markets can lead to inefficient production levels if individuals and firms do not have common information. If some individual or firm is ill-informed relative to another, it may find it difficult to profit from exchange with the better-informed party, and there may be no exchange at all. A fruit grower knows what pesticides he applied to his product, but if the retailer and the consumer do not, then they may have to take costly measures to reduce the risk of illness; alternatively, the grower and the retailer may find it costly to communicate information about pesticides to consumers in a useful way. Government intervention may take the form of food safety regulations.[2]

The public good character of primary and secondary data, and analyses based on these data, can lead to asymmetric information when there are many small participants on one side of a market for a commodity and a few large ones on the other. Each large participant realizes some return from investing in data collection and analysis, but on the other side of the market there is negligible return from these activities for any one participant. In this situation, the level of production of the commodity is very likely to be too high or too low. Government intervention to provide data and analyses can then bring about a more nearly optimal level of production of the commodity. Political support for this intervention may be grounded in considerations of equity, as has been the case historically in many agricultural markets.

Property Rights

Markets function only in the context of well-defined property rights. As technology changes, issues of property rights continually emerge, and a primary function of government is to establish property rights appropriate to the state of technology. For example, advances in molecular biology have greatly accelerated the development of new strains of crops, including those resistant to disease and infestation. What are the intellectual property rights associated with the new strains? Advances in communication, including cellular telephones, cable television, electronic mail, and Internet services, have greatly multiplied the uses of the electromagnetic spectrum. In view of these changes, what form should media ownership take? As the impacts of industry and agriculture on the environment become better understood, new kinds of property—for example, an upper atmosphere undepleted of ozone—take on value, and questions of ownership come to the fore. Only governments can provide the constantly changing infrastructure of property rights essential to the efficient functioning of private markets.

---

[2]Of course, fruit producers, retailers, and consumers could be equally uninformed about the safety of pesticides. New knowledge about pesticides is a public good. With more than 10,000 fruit producers, more than 1 million food retailers and eating establishments, and more than 100 million consuming households, new knowledge is nonexcludable as well and therefore a pure public good. Thus it is quite likely that the market produces too little information about the impact of pesticides, if any, on consumer well-being. So there is a strong public good argument for food safety regulations, too.

Equity

Governments often intervene in markets deliberately to change the distribution of income and wealth. The intervention may be motivated by broad consensus about an appropriately equitable distribution of income, or it may reflect the give and take of particular interests with respect to specific issues. But the distribution of income is invariably changed by any government economic intervention. Understanding, much less accurately anticipating, the distributional and other effects of economic interventions is a difficult undertaking, even when conducted by a disinterested and skilled party. For example, crop price support programs have had their ultimate impact on the value of land, not on the earned incomes of farmers, and at the same time they have adversely affected soil conservation by creating disincentives for crop rotation. The greater is the uncertainty about the distributional implications of existing and proposed interventions, the more likely it is that specific groups will claim or fear widely varying consequences.

Other Factors

Government economic intervention is undertaken for a wide variety of other policy purposes, as well. For example, in the United States as well as other industrialized countries, there are long-standing programs to maintain or improve the welfare of rural populations. In an earlier era, programs for this purpose included rural free delivery and rural electrification in the United States. Today, these programs include measures to increase the flow of information to rural areas, for example through Internet access, and federal partnerships with nonprofit organizations providing assistance to rural areas.

WHY GOVERNMENT-SPONSORED RESEARCH?

Basic Research

Government supports many research programs. In 1997, federal support for nondefense research and development was $28 billion (National Science Foundation, 1998a). Most of this expenditure is in basic and applied physical and biological sciences. Some federal agencies, for example the Agricultural Research Service of USDA, conduct this sort of research primarily within the agency; others, like the National Science Foundation, primarily sponsor research in the academic sector. This is an important economic intervention. It occurs primarily because the outcome of basic scientific research is usually a pure public good; the outcome of applied scientific research can be, as well. The key economic property of the research outcome is whether or not it is excludable. Breakthroughs in basic mathematics and the development of an improved surgical technique are examples of nonexcludable public goods. The discovery of a new drug and the design of a new computer chip are public goods, but they are excludable through the patent system. Given the large number of corn producers in the United States, basic development of hybrid corn earlier in this century was a pure public good, and it was not until the 1930s that replication of inbred lines and distribution became excludable and thus primarily a function of private-sector firms. A vertically integrated producer of poultry with sales in the billions of dollars will have incentives to undertake both basic development and subsequent marketing of a patented, excludable commodity.

Government-sponsored data collection began with the constitutionally mandated decennial census in 1790, and the earliest collection of agricultural data by the federal government was undertaken by the Patent Office in 1840. Data are manifestly public goods: use of data by one party in no way diminishes its usefulness to other parties. They need not be pure public goods, because they can be excluded. For example, econometric consulting firms provide data in electronic form under licensing agreements that prohibit further disclosure of the data. This excludability of data, a comparatively recent development, was much costlier in the nineteenth century when USDA began systematic data collection. Similarly, if data are collected by literally going door to door, it is more efficient for one collector to gather all the information from that door, than for many collectors to go to the same door. With electronic communication, there can be many data collectors, and the case for a natural monopoly weakens. Data collection by a disinterested party can ensure that all parties have access to information in a way that collection by an interested party cannot. This does not necessarily imply that government must intervene. In many industries, the best data are collected by parties under contract to industry-wide associations. In some cases, firms in the industry do not trust each other or their association, and data collection is contracted to a trusted government agency.

Data collection at USDA evolved naturally into monitoring and reporting, as detailed in Chapter 3. The line between data and the interpretation of data is subjective, but by the 1920s the BAE was providing forecasts of commodity prices. Monitoring and reporting began at a time when, by comparison with the present day, information was very limited, sophisticated financial markets for the purpose of conveying information scarcely existed, and there was widespread misunderstanding of the role of information in a modern economy.[3] The case for nonexcludability of monitoring and reporting, as well as an asymmetric, informational disadvantage for farmers, was stronger in the 1920s than it is today. Government agencies today collect data extensively, but the monitoring and reporting function is often left to the private sector. For example, the Forest Service of USDA sells timber tracts at auction and keeps detailed records of sales and subse-

---

[3]For example, cotton price forecasting by USDA was proscribed by Congress in 1929. This legislation still stands.

quent logging of the tracts. But it does not provide the same sort of monitoring and reporting that ERS provides in its *Situation and Outlook* reports.[4] That function is fulfilled by a private firm that sells a detailed newsletter to mills in the industry.[5]

Why Research in Economics?

The past two centuries have seen by far the greatest advancement of living standards in human history. The development of new technology has been essential for these advances, but technology alone has not been sufficient. The invention of the steam engine had its effect on living standards through the development of new industrial processes and rail and sea transportation, all of which required much more than technological advances. The economic success of rail transportation, for example, also required innovations in property rights, the development of a large supporting economic infrastructure, and eventually substantial modifications of the market economy itself. The history of these changes can be read in many Supreme Court decisions of the late nineteenth century. A generation later, a similar process of institutional change accompanied the economic impact of the automobile. The process is being repeated today with innovations in information technology. Realizing the full potential of these innovations is entailing changes in work habits, in definitions of intellectual property, and in the education system.

The changes that must follow scientific and engineering innovations for these advances to contribute to human well-being are primarily economic changes, broadly defined. The need for economic knowledge grows out of the need for institutional change and improvements in institutional performance, driven by technical progress. The knowledge required includes definitions of property rights and an understanding of the ways in which markets will succeed and will fail—in the ways just considered—as society accommodates and exploits new technical knowledge. Technical knowledge, and its application, are becoming more and more widely available: software is written in India, shoe manufacturing takes place in Malaysia, and light aircraft are produced in Brazil. In a world of freer trade and shared information, the relative success of nations depends as much on the successful adaptation of their economic systems to technical innovations as it does on the output of their research laboratories (Porter, 1990).

The systematic, analytical study of the economic aspects of scientific progress has been essential to the rapid advancement of American living standards in the twentieth century (Ruttan, 1984:552):

---

[4]These reports are described at the start of Chapter 5.

[5]Government provision of information to rural areas could be undertaken in an effort to subsidize and thereby maintain small rural businesses and thus rural populations. In such a situation, this kind of information provision might be compared with other subsidies for the same purpose, for example, tax incentives for firms to locate in rural areas.

Throughout history, improvements in institutional performance have occurred primarily through the slow accumulation of successful precedent, or as a by-product of expertise and experience. Institutional change was traditionally generated through the process of trial and error much in the same manner that technical change was generated prior to the invention of the research university, the agricultural experiment station, or the industrial research laboratory. With the institutionalization of research in the social sciences it is becoming increasingly possible to substitute social science knowledge and analytical skill for the more expensive process of learning by trial and error.

The contributions of economic research are to identify the changes in institutions and physical and human capital necessary to exploit new technology and to identify the effects of these changes. It offers the opportunity to suggest the changes that are most likely to succeed, and thereby to reduce the costs of institutional innovation. Changes in institutions include modifications of government policy that address market failures and inequities in income distribution, as well as property rights and other aspects of the political infrastructure in which a market economy functions.

A particularly important contribution of economic research in a representative democracy is to identify in some detail the effects of changes in policy (or, for that matter, the effects of leaving policy unchanged in the face of changing technology) on different economic interests. It is rarely, if ever, the case that a change in policy will leave every interest in an improved or equivalent condition. But if those who gain do so enough that those who lose can in turn be compensated in such a way that no party loses ground, then a combination of a change in policy and redistribution may be politically feasible. The identification of potential winners and losers *beforehand* is a contribution of economic research that can facilitate the political process.

Why Public-Sector Research in Economics?

The need for economic knowledge, driven by change in technology, is both private and public. Private firms allocate very substantial resources to the acquisition of economic knowledge in the pursuit of economic efficiency. This is especially the case for large enterprises, for example multinational firms that deal in a variety of legal and institutional environments. Markets for information have become extremely important components of the world economy, including not only securities but also derivatives of securities like futures and options contracts.

Addressing Difficult Issues

The largest and most difficult questions remain public, however. In a representative democracy, the establishment of property rights is inherently a public issue. Problems of market failure from environmental externalities alone demand increasing public attention. Issues of income distribution are as important today as they have ever been. Economic knowledge has been developed and applied in addressing many of these questions. For example: the creation of property rights and subsequent allocation of part of the electromagnetic spectrum has made use of research on auction design (McAfee and McMillan, 1996). The institution of time-of-use pricing to delay the construction of new electricity-generating capacity has drawn on extensive controlled experiments carried out by econometricians (Aigner, 1981). Concepts of property rights have been extended and new market mechanisms have been developed as policy makers have balanced the amelioration of negative environmental externalities with economic efficiency. The creation of a market for sulfur dioxide emission licenses for electric power plants is one example. Another is the Conservation Reserve Program designed largely by ERS, in which farmers bid to take environmentally sensitive acreage out of production for 10-year periods.

Use of public economic research by one organization or individual does not diminish its value to anyone else. It is nonrival and therefore a public good. Many kinds of public economic research are also nonexcludable and are therefore pure public goods. For example, advances in computable general equilibrium models have made more timely and accurate anticipation of changes in taxes and tariffs possible, but new ideas as such cannot be patented, and these advances are therefore nonexcludable. Private markets will underproduce economic research because of its pure public good character. Economic information is also nonrival, but it may be excludable. For example, ERS regularly provides information of keen interest to various industries, as do industry newsletters, but given current technology this information is often excludable (for example, by limiting electronic access).

Uses of Private-Sector Research

In apparent contradiction, there is widespread production of economic knowledge by private interests on many economic questions. Any contemplated important change in public economic policy is likely to bring forth a host of studies originating in the private sector. On one hand, it is vital for private-sector interests to be able to identify impacts of contemplated policy change that may have gone unnoticed. On the other hand, many of the effects identified by private-sector interests are rent seeking: that is, a private-sector interest that stands to benefit substantially from a proposed change may support that change with evidence indicating that it will benefit some wider group. The claim may or may not be true, but in this situation the private-sector interest cannot be expected to provide evidence on who stands to lose.

The outcome of this sort of private-sector research is well understood and well anticipated by a simple analysis of market failure. To the extent that the gains or losses of a proposed policy change are spread out, with no single interest

SOWING SEEDS OF CHANGE

much affected, no single interest will have an incentive to produce the economic knowledge. To the extent that the gains or losses are concentrated on a small collection of interests, the incentive to produce the knowledge increases. This situation has at least two very undesirable features. The first is that policy changes tend to reflect concentrated interests. The second is that this fact may emerge only well after the change has been made, if at all, in an environment of ongoing changes in diverse economic policies. This is not even policy experimentation: it is policy chaos.

Advantage of Prior Analyses

Public-sector research in economics complements but does not displace private research. Gathering evidence on likely outcomes before the fact, rather than after, affords the potential for large gains. First and most important, it provides the only alternative to carrying out the field experiment of actually making the policy change. For example: research on auction design is a very attractive alternative to uninformed experimentation with the rules for government auctions, for example in the sale of the electromagnetic spectrum and the leasing of productive cropland for conservation purposes. Controlled experiments in time-of-use pricing for electricity on a small scale indicated the pricing schedules that would best achieve energy conservation and environmental goals on the large scale. And the work of ERS in the 1980s provided the basis for prior assessment of important aspects of current world trade agreements on agriculture, whereas experimentation with actual policy change would have been extremely expensive.

Second, studying outcomes before rather than after the fact can identify gainers, losers, and the potential for redistribution from the former to the latter, thereby providing sound information from which essential political agreements can be struck. For example, ERS demand modeling made an important contribution in support of the Uruguay round (1986-1994) of trade negotiations under the General Agreement on Tariffs and Trade (GATT): it identified the important domestic gainers and losers in the principal agricultural exporting nations. This enabled policy officials to convey to Congress that domestic political interests who opposed the agreement were overstating their potential losses, and to make the case that the losses of losers from liberalized agricultural trade would be more than offset by gains of the winners.

A third attraction of economic analysis before the fact is that it suggests what should be monitored after the fact in order to evaluate the impact of the change in policy. This last contribution remains intact even if public-sector research on the issue is ultimately ignored in reaching the decision about policy change. The evidence gathered subsequently may have bearing on future policy questions and can be used to compare the consequences of policy change with predictions before the fact, thereby improving economic analysis. For example, ERS has long monitored research expenditures and measures of farm productivity to facilitate the assessment of returns to research in agriculture. It maintains information on agricultural output and prices and models of supply, which have been used to identify the effects of proposed changes in subsidy programs during the consideration of major farm legislation that must be renewed about every five years.

Two Examples

Economic research is carried out in academic institutions, in private firms, and in government agencies (primarily federal). The product of research activities is new knowledge. A straightforward consideration of the market for this new knowledge typically explains the kinds of economic research conducted in these three sectors, as the following two examples illustrate.

Many assets, goods, and services are bought or sold by means of auctions. This is especially the case for assets, goods, and services acquired or divested by governments. Auctions can be conducted in a wide variety of ways: bids can be oral or written, the price in an oral auction may be ascending or descending, the price paid may or may not correspond to a highest or lowest bid, the seller or buyer may or may not announce a minimum or maximum acceptable price. To a buyer or a seller organizing an auction, the most important question is how to minimize or maximize (respectively) the price of the object being bought or sold.

The relation between the organization of the auction and the transaction price of the object depends on economic characteristics of the object and the potential buyers and sellers that transcend specific settings. Knowledge about this relation is a public good, because it is not diminished by its use, and it would be difficult to both apply this knowledge and exclude it from others. Thus it is not surprising that advances in the theory of auction design (Engelbrecht-Wiggans et al., 1983; Hirshleifer and Riley, 1992) have taken place almost entirely in the public sector. The theory identifies the economically relevant characteristics of an auction that in turn predict the consequences of alternative auction designs. Ascertaining these characteristics for a particular object being sold may or may not be a public good. For example, procurement of milk for public school lunches is a similar process in most states and school districts. Determination of the economically relevant characteristics is a public good, and no one school district has much incentive to carry out this research. Most of the work has been done by academics. In contrast, the features of the market for procurement of weapons by the U.S. Department of Defense are unique to that market, and the department has sponsored research to study these features and improve the design of its procurement auctions.

A second example is the research that has provided the basis for time-of-use pricing of electricity. Because the generation and delivery of electricity has been a natural monopoly, it has been regulated by a public utility commission in each state. Demand for electricity varies systematically throughout the day and throughout the year. Since electricity cannot be stored efficiently, generating capacity must be adequate to meet the maximum rather than the average demand for its use. However, by changing the price of electricity systematically throughout the day or the year, it is possible to alter the systematic variation in demand. Recognizing the advantages of studying policy changes before rather than after the fact (Aigner, 1981), the (then) Federal Energy Administration supported a series of experiments to quantify the relationship between the pricing policy and the demand for electricity. Constructing the econometric framework and sampling design for these studies raised similar methodological issues across states. Research addressing these issues was carried out predominately by academics, with findings published in research journals and therefore widely available. In applying the econometric framework and sampling design, issues specific to each state arose, since industry mix and seasonal demand for electricity vary widely by state (Aigner, 1981). This work was sponsored by individual state utility commissions, typically by contract to individuals or private-sector research organizations.

In both of these examples, the basic research produced knowledge that is close to a pure public good. The economics of auction mechanism design and the econometrics of electricity pricing experiments are in no way diminished when they are applied to yet another policy problem, and it is essentially impossible to exclude this knowledge from use by others. Indeed, centuries of experience have established that basic research is most successful when it is publicly available, so that those who do the same kind of research can criticize it and build on it in a timely fashion. Economic and many other kinds of research in the United States is based on a partnership between the federal government and academic institutions, one that explicitly recognizes the public good nature of basic and applied research and is designed to produce it efficiently.

This partnership is founded on two secondary marketplaces. In one, academic advancement and salaries are based on the contribution to knowledge as measured by peer evaluation, largely through the medium of peer-reviewed scientific journals. In the other, resources for carrying out research are provided by the federal government based on the prospects for their efficient use as measured by past performance and peer evaluation of proposals. The two secondary marketplaces are closely related. They provide a very strong set of incentives for creative, productive work that has made the United States the leader among all nations in economic and other research. It brings many of the brightest young scientists to the United States for training, and many of the best of these stay on and make further vital contributions.

In both the auction design and electricity pricing examples, the basic research conducted in the public-academic sector was used in private decisions as well as in the making of public policy. In both cases, the decision makers were aware that new research coming out of the public-academic sector could contribute to a better decision. They determined the bearing of this research on the decision they faced, incorporating it into the decision itself. Establishing the link between basic research, on one hand, and a decision, on the other, is a critical step that typically demands substantial time, talent, and resources. If the link has previously been made by others in similar situations, then it is inefficient to repeat the process from the ground up each time. This gives rise to extensive markets in applied research in support of decision making, involving both private- and public-sector entities. In the private sector, it includes consulting firms and newsletters. In the public sector, it includes university professional schools and some of the activities of community colleges and vocational schools.

Effective Economic Policy Analysis

More novel circumstances are likely to demand greater sophistication in bringing economic knowledge to bear on the question at hand. To the extent that a situation is novel, the simple model of imitating others in similar situations is more likely to lead to a bad outcome, and the more critical it becomes to establish a solid economic basis for the decision. In the private sector, novel decisions are faced by firms in new industries, and markets provide substantial rewards for making such decisions effectively. In the public sector, decisions in novel circumstances must be taken by governments confronting the institutional aspects of technical progress for the first time. The U.S. federal government is in this position more often than any other public institution. But there is no market that provides timely rewards for the effective application of knowledge in these critical decisions. The only available substitute is a public policy that effectively links basic economic research to decision making.

The link between research and economic policy decisions is *economic policy analysis: the disinterested prospective and retrospective evaluation of the economic and social implications of changes in public policy, and the effective communication of these evaluations to public policy makers.* Effective policy analysis requires understanding proposed changes in public policy and the problems that drive those proposals, in order to frame the implications of changes as questions that have been, or can be, addressed in basic and applied economic research. At the level of federal policy, it is unlikely that the proposal at hand will have been addressed in the best way by previous applied research; in any event, knowing the body of applied research that comes closest is a demanding profession. Policy analysis does not emerge as the product or by-product of basic and applied research elsewhere or in economic research in support of private decisions. Effective and therefore timely policy analysis cannot be delegated to the academic sector, and evaluation that is both disinterested and relevant will not be carried out in the private sector.

For public policy support research to maintain its promise of reliably identifying the likely effects of changes in policy and institutions, it must have four characteristics. It must be of *high quality*. By definition, research of higher quality provides a more reliable assessment of change than research of lower quality. Research that does not meet disciplinary and professional standards, or is grounded in poor or inappropriate data, runs the risk of being dismissed in favor of higher-quality research, including research carried out by stakeholders in the policy question at hand. Public policy support research must be *relevant*. Research that does not address the essential policy question will either be dismissed by policy makers or, in addressing the wrong question, may provide the wrong answer. It must be *timely*. Research undertaken prospectively may avoid the costs of social experimentation. Moreover, research that is conducted in anticipation of policy questions and concluded before political lines are drawn is more clearly not beholden to specific interests than is research that is carried out after positions, including the position of a cabinet secretary, have been established. Public policy research must be *credible*. This characteristic derives from the first three. Any entity that regularly produces high-quality, relevant, timely research will enjoy a reputation for credibility, as well as independence from narrow parochial or political interests.

These characteristics can be maintained only if the distinction between policy analysis and the decision itself is maintained. Ideally, research in support of public policy sponsored by a federal agency will be the same, regardless of the political position of the president or the secretary, whereas the way in which that analysis is used will depend very much on those officials' political and policy positions. As Chapter 3 reveals, the history of the BAE and ERS amply demonstrates how destructive direct involvement in the political, problem-solving process is to policy analysis and its promise of improving public decisions. If policy analysis survives at all in this environment, then it is no more credible than the research of private interest groups. The more likely outcome is the departure of key members of the research group and its subsequent decline or dissipation. The decline of BAE prior to its dissolution in 1953 documents this consequence of mixing research and policy decisions or implementation. In the short run, the forces that would bend the factual outcome of policy analysis to political purposes are strong, but, in the long run, yielding to these forces renders policy analysis useless and policy decisions less well informed. In the long-term, open and credible policy analysis supports good decision making, can be a potent political and positive force, and can generate the political support necessary to protect and sustain credible research and analysis in government.

What Should We Expect from the Research Process?

The outcome of a research project is never clear when the project begins. The less routine is the problem addressed, the greater is the uncertainty about the outcome. Many research projects will have disappointing outcomes: new basic research may not exhibit the promise it seemed to hold; a key conjecture may turn out to be right but of little use in addressing the question; a new data set may turn out to have the same limitations as a data set previously available. But in a smaller set of cases, the payoff may turn out to be high. For example, liberalization of agricultural trade in the Uruguay round of trade negotiations under the GATT agreement is estimated to have net benefits of up to $8 billion annually for the United States and $30 billion annually worldwide (Economic Research Service, 1989:Table 10). ERS had a small but vital role to play in the negotiations, and its annual budget is a very small fraction of the net benefits realized from the new agreement. Policy analysis, like research, shares an important characteristic with drilling for oil: failures often outnumber successes, but the gains from the successes exceed the costs of the failures.

For research in support of public policy, there are two further apparent risks. The first stems from the need to anticipate policy problems. In some instances, the time horizon for policy questions is clear. For example, the next round of World Trade Organization negotiations will begin in 1999, and trade issues in agriculture will be on the agenda. In other instances, it is not: hog waste may remain a local environmental problem, or it may emerge as a policy question at the federal level. It is inevitable that a group formed for public policy research support will anticipate some questions for which it is never asked to have problem-solving research available. Research in support of public policy is somewhat like investment in defense, made in anticipation of many contingencies, most of which may never arise. Sponsors of such efforts must accept this as a cost of the other successful applications of research.

Second, even an important, relevant contribution of policy analysis may not be reflected in the final policy decision. The decision is the outcome of a political process, and those who engage in policy support research are well aware that a proposed policy innovation well grounded in excellent research may not survive an eleventh-hour compromise.

These two features are risks from the narrow perspective of evaluating the utility of research with respect to the question that motivated it, but they are less risky from the appropriately broader perspective of national policy. Much of the research undertaken for policy support at the federal level will also provide useful support at the state and local levels, precisely because the questions it addresses emerge at those lower levels of government rather than as national policy issues. Indeed, the public good characteristics of research argue strongly for a federal role in support of problem-solving research even when used only by states and localities. The argument is all the more compelling with the devolution of federal programs to the states. Problem-solving research that ultimately does not affect policy is often still useful in retrospective evaluations; if the policy turns out to have addressed the problem poorly, then its day may come again.

3

The Lessons of History

The Economic Research Service (ERS) and its predecessor, the Bureau of Agricultural Economics (BAE), have performed in a variety of roles over the three-quarters of a century since the BAE was established in 1922. Much can be learned from this long experience about the limitations and the imperatives facing a government agency responsible for research, secondary data development, and various types of analysis.

The reporting structure and organizational environment of ERS and the BAE have changed greatly over time, with substantial implications for the role, political limitations, and expected performance of a research, information, and analysis agency and for its internal culture, level of professionalism, and quality of product. It is also the case that, over the past 75 years, the nature of the food system, and consequently the U.S. Department of Agriculture (USDA) political and policy agenda, have changed drastically. These changes have created a progression of rather different challenges for the provision of objective information, research, and analysis, which suggest some of the specific goals, strategies, and behaviors important to such an agency's success. This chapter examines the experience of the BAE and ERS in order to extract the lessons history provides for a primarily economics-based federal agency producing research, information, and analysis.

ECONOMIC INTELLIGENCE AND RESEARCH IN USDA

Although some agricultural data have been collected by the government (the Patent Office) since the 1840s, the Department of Agriculture, which was founded in 1862, created a Division of Statistics in 1863 to take over this responsibility.

USDA received its first appropriation specifically for collecting agricultural statistics in 1866 (Duncan and Shelton, 1978:6; Tenny, 1947:1018). Since that time, regular monthly reports have been produced on conditions of crops and numbers of livestock on farms. The purpose of this public intelligence was to improve the equity and efficiency of local commodity markets and to warn of disease, droughts, and other local farm problems. In the nineteenth century, this became a progressively larger challenge due to the ever-growing expanse and biodiversity of farming, as the U.S. frontier marched across a continent. Better-informed decisions by the millions of farmers and thousands of agricultural market firms were the immediate objective.

Public policy in agriculture—and thus the mission of USDA—was originally limited to gathering and disseminating market intelligence and to natural science research and education designed to improve agricultural productivity and defend it against disease and pests. Throughout the nineteenth century, USDA was primarily a statistics, research, and education organization. Its likeness to a university was often remarked upon (Gaus and Wolcott, 1940:16; Wilson, 1912).

Regulation of agricultural markets also began in the late 19th century—primarily for animal health and food safety purposes and to establish common standards for weights and measures. Direct public policy intervention in markets to affect farm prices and incomes, i.e., the farm programs, did not occur until the farm crisis of the 1920s and the Great Depression of the 1930s. Social science, primarily economic research, began to develop as part of the USDA's mission only in the first decade of the twentieth century (Baker et al., 1963).

The first formal organization for data collection in USDA started with the establishment of the Division of Statistics in 1863, one year after creation of the Department of Agriculture itself. Agricultural products constituted well over half, and in some periods more than 80 percent, of all U.S. exports during the nineteenth century (Cochrane, 1993:150, 267-270; Economic Research Service, 1977). By 1902, a Division of Foreign Markets had been established to develop data on world agricultural markets (Baker et al., 1963:517). In 1903, the Foreign Markets Division was merged with the Division of Statistics to form a Bureau of Statistics. In 1914, the Bureau of Statistics was renamed the Bureau of Crop Estimates, and in 1921 it was merged with the Bureau of Markets to form a Bureau of Markets and Crop Estimates (Baker et al., 1963:80, 516-517). This merger brought together, in one organization, responsibility for the collection of farm-level crop and livestock data with that for major domestic and foreign commodity market transactions.

At the same time, economic research was slowly being developed in USDA and focused on the problems and decision needs in farming. The Office of Farm Management had first been formed in 1903 in the Bureau of Plant Industry, but it was transferred to the Office of the Secretary in 1915 (for economic and program analytic support during World War I). Its title changed to the Office of Farm Management and Farm Economics in 1919, and then in 1920 it was moved out of the secretary's office to separate USDA agency status (Baker et al., 1963:107-108, 498-501). Finally, in 1922, the data collection function of the Bureau of Markets and Crop Estimates was combined with the economic analytic and research capacity of the Office of Farm Management and Farm Economics to form the Bureau of Agricultural Economics. Most BAE economists were recruited from a few land grant colleges of agriculture with Ph.D. programs. Even as late as 1930, the vast majority of U.S. agricultural economists were being trained at only three universities: the University of Wisconsin, the University of Minnesota, and Cornell University (American Farm Economics Association, 1930).

As farm production became less and less a matter of subsistence during the nineteenth and early twentieth centuries and more a monetized market transaction, the dependence of millions of farmers and market firms on USDA for market information grew apace. Over this period, farm markets expanded from primarily local or regional to national and, for some commodities, even to international in scope. At the same time, the nature and scope of the demand for market intelligence, economic research, and analysis also expanded. Periodic recessions and banking "panics" led to disordered farm markets and major declines in farm prices and welfare during the nineteenth century, periodically demonstrating the importance to the efficiency of farm markets of broad public access to more complete and accurate market intelligence. By the turn of the century, concern for the welfare of farmers, most of whom were quite poor, led to research to understand and deal with the now-chronic economic problems of an industrializing agricultural sector that exhibited immense and repeated price and income instability (Taylor and Taylor, 1952:1-53; Edwards, 1940; Davis, 1940).

During the late nineteenth century, an academic field of farm management teaching and research had begun to develop in the land grant colleges in response to the growing complexity of farmers' management decisions under the impact of nearly continuous technological change and the commercialization of farm production. The earliest pioneers were typically agronomists and horticulturists, such as Thomas F. Hunt and Liberty Hyde Bailey at Cornell, who taught some of the first courses in the economics of agriculture. Farm management eventually evolved into today's agricultural economics profession, members of which provided much of the early leadership in the development of the Bureau of Agricultural Economics and its precursor organizations (Taylor and Taylor, 1952:53-156). By 1910, USDA was generally considered the finest agriculture science research organization in the world. The BAE reached equivalent status in agricultural economics in the 1930s and 1940s.

The importance of economics in agriculture and the professional development of agricultural economics is primarily associated with four names. H.C. Taylor at the University of Wisconsin was most influential over the period 1890 to 1930. George F. Warren, of Cornell University, was a major influence from 1900 into the 1930s. John D. Black, while at the University of Minnesota and later at Harvard University, was active from the 1920s through the 1950s.

Theodore W. Schultz, first at Iowa State College and then the University of Chicago, profoundly influenced the discipline from the 1930s to the 1980s. Taylor, Black, and Schultz earned their Ph.D.s at the University of Wisconsin, Schultz and Black both under the direction of Benjamin H. Hibbard. There were many others who contributed, but these four and their many students had a profound influence on the development of the economics of agriculture. This group and Taylor's students at Wisconsin were omnipresent in the development and direction of the BAE (Gilbert and Baker, 1997).

The problem faced by agricultural economists in the 1920s and the 1930s was a severely limited data base for analysis and an underdeveloped conceptual framework. The long-term effort in agricultural economics and the BAE was to develop a stronger data base to support analysis, along with greater sophistication in quantitative measurement and improvement of the economic and other conceptual foundations of their work. Their success was attested to in a presidential address to the American Economic Association in 1970. In it, Harvard professor Wassily Leontief (and 1973 Nobel laureate) indicted the economics profession for generally failing to design and collect data for the empirical tests necessary to validate the developing theoretical base of economics. In the speech, he exempted agricultural economics from his indictment as having achieved "an exceptional example of a healthy balance between theoretical and empirical analysis and of the readiness of professional economists to cooperate with experts in neighboring fields" (Leontief, 1971:5). Much of this early achievement can be credited to the BAE. From the late 1920s and the 1930s onward, many of the intellectual leaders of the agricultural economics and statistics professions worked in the BAE.

THE BAE AND LESSONS LEARNED

The BAE was established in a period of growing economic distress and conflict in agriculture. The United States had expanded farm production during World War I to feed a war-ravaged Europe. Following the recovery of European production after the war, U.S. farm prices and land values collapsed in 1920-1921, creating another major economic crisis in U.S. agriculture. U.S. farmers were left with heavy debts and excess capacity. Bankruptcy became endemic. Serious economic problems in farming continued through the 1920s into the Great Depression of the 1930s (Davis, 1940). Restrictive macroeconomic and trade policies compounded the world's economic problems. This intensified the demand not only for economic intelligence but also for economic research and policy analysis to inform public and private decisions dealing with the crisis in the United States.

From its establishment in 1922, the BAE found itself in the middle of intense political debates regarding farm relief and appropriate farm policies through the 1920s and the 1930s. This period of unending philosophical conflict over values and beliefs in politics and policy is well described by two early BAE leaders, H.C. Taylor (1992) and Nils Olsen (Lowett), as well as by historian Richard Kirkendall and political scientist Charles Hardin.

The politicized atmosphere and ideological conflict led many interests to support and many others to question the worth of the BAE's statistical and analytical work, which was growing increasingly sophisticated and complex. At the same time that the BAE sought to standardize and quantify its analytic methods, it found itself under public attack; farmers and their organizations often questioned the validity of BAE estimates, which they found difficult to comprehend and the results of which often conflicted with their current perceived policy interests (Kunze, 1991:79).

As a consequence, many politicians, especially some members of Congress, grew hostile to BAE crop estimates, since they believed it was the estimates and associated price forecasts (not macroeconomic policies and declining demand or increasing supply) that were causing declines in farm prices. Congress reacted in 1929 by passing a law prohibiting the USDA from publishing any cotton price forecasts—and this legislative prohibition still stands (Townsend, 1987; Black, 1953:386-387). In addition, during the Hoover administration (1929-1933), the prohibition was extended to other farm commodities by executive action. BAE forecasts were limited by the administration to those with a "need to know" and could not be provided to the general public (Kunze, 1991:79). Faced with a debilitating problem of economic illiteracy amongst farmers, the BAE in cooperation with the land grant colleges early put an expanding effort into economic education, especially in conjunction with and in support of the market information (i.e., *Situation and Outlook*) work begun in the 1920s (Taylor and Taylor, 1952:447-479). The BAE outlook program expanded previous USDA market intelligence and added economic analysis. Political opportunism combined with very low levels of economic literacy among farmers, organized interests, federal program agencies, and in politics have historically plagued USDA analysis of farm markets (Cochrane, 1965:456-457; 1983:87; Lowett, 1980:65-66).

Secretary of Agriculture David F. Houston (1913-1920), an economist by training, presided over and supported the creation and development of economic work in USDA. H.C. Taylor was head of the Department of Agricultural Economics at the University of Wisconsin in 1919, when Secretary Houston invited him to Washington to head the Office of Farm Management and to help organize the economic analytic capacities of USDA. The next secretary, Henry C. Wallace (1921-1924), strongly supported the developing economic work of the department. During Wallace's tenure, a period of considerable economic turmoil in agriculture, the BAE was created and Taylor became its first chief (Baker and Rasmussen, 1975:55; Taylor 1992:31-41).

The post World War I crisis in farming was generating widely differing proposed solutions to the "farm problem." Not surprisingly, this led to intensely partisan political conflict in the 1920s and the early 1930s. The different "solutions" ranged widely: letting the markets work without government intervention, government-led export dumping, turning management of markets over to farmer cooperatives to control production and thus prices, government-guaranteed purchasing-power parity for farm prices, and a two-price domestic allotment plan to separate the domestic market from foreign markets (thus protecting a higher subsidized price for domestic producers). The BAE was asked by the various secretaries of agriculture to evaluate different proposals, some of which were unworkable as presented and all of which involved costs or problems that advocates did not want exposed in public forums. In the midst of this political turmoil in the 1920s and the 1930s, the BAE under Taylor attempted to provide objective research and policy analysis. This objective analysis attracted critics and left the BAE and its leadership quite vulnerable when not protected by the administration.

After Secretary Wallace's death, Secretaries Gore (1924-1925) and Jardine (1925-1929) protected BAE Chief Taylor for a while, but Jardine, unable to convince him to resign, finally fired Taylor in August 1925 at the Coolidge White House's insistence (Baker and Rasmussen, 1975:55; Taylor, 1992:207-219). BAE leadership continued to turn over rapidly during the next few years. The fourth chief of the BAE, Nils Olsen, survived seven years (1928-1935) under three secretaries by following a conservative, risk-averse strategy that tried to serve the secretary's needs but kept himself and the BAE away from public involvement in analysis of policy issues on which the administration had not yet taken, or was unwilling to take, a position.

Olsen was fully committed to an objective research and information role for the BAE, but he was not a supporter of the intrusive New Deal programs. Eventually Olsen resigned in 1935 when, in his view, the new secretary, Henry A. Wallace (1933-1940), who was the son of Henry C. Wallace, grew less and less supportive. Wallace was reluctant to make decisions in an unstable political and policy environment that had, in the early New Deal, dissolved into unending ideological warfare (Lowett, 1980:222-235; Baker and Rasmussen, 1975:56-58; Kirkendall, 1982:16-17, 77). Secretary Wallace, despite his cool relations with Olsen, was an astute and heavy user of BAE intelligence and research. Indeed, beside his success in creating the first hybrid seed company and as editor of *Wallace's Farmer*, he was an accomplished professional geneticist, statistician, and economist.

Transformation of the USDA Role

Between 1933 and 1939, the nature of USDA was fundamentally transformed. New Deal programs of the Roosevelt administration (1933-1945) addressed the multiple problems of agriculture and rural America and led to the creation of numerous action agencies. Created in this period were the Soil Conservation Service, 1935 (now the National Resource Conservation Service); the

Rural Electrification Administration, 1939 (now the Rural Utilities Service); the Foreign Agricultural Service, 1938; the Farm Security Administration, 1937 (later renamed the Farmer's Home Administration and today divided between the Farm Service Agency and the Rural Housing and Community Development Service); the Federal Crop Insurance Corporation, 1938 (now part of the Farm Service Agency); the Commodity Credit Corporation, 1933; and the Agricultural Adjustment Administration, 1933 (known then as the AAA or Triple A, later renamed the Agricultural Stabilization and Conservation Service and now part of the Farm Service Agency) (Baker et al., 1963:463-519). These organizations delivered services or resources or both to farmers and in many cases developed offices at the state and local level with farmer advisory committees that quickly became specialized clientele organizations. They lobbied Congress and USDA in support of the agency and its program (Kirkendall, 1982:167-223; Lowi, 1962).

This approach fragmented—and focused by function and commodity—the political interests in agriculture (Lowi, 1962; Heinz, 1970). The larger interests of the food and agricultural industry and of the nation tended to be ignored, except as the secretary and the president imposed them on the debate. The nature and dynamics of agricultural politics changed in Congress, the secretary's office, and the countryside. The internal environment of USDA was soon filled with politically powerful interests that made the secretary's role far more difficult and outmatched the political influence of all of the research and education organizations that had characterized the earlier USDA (Hathaway, 1963:201-206, 210-214; Gaus and Wolcott, 1940:64-81, 264-265; Rasmussen and Baker, 1972:Chapter 3; Lowi, 1962; Truman, 1965). In the external environment of the department, the development of a state and local grassroots program and political presence for many of these federal action agencies created a competitive, or potentially competitive, relationship with the politically active farmer organizations and with the land grant colleges' Cooperative Extension Service (Hardin, 1952, Chapters 6, 9, 15; Lowi, 1962; Rasmussen and Baker, 1972:Chapter 12). Thus, not only had the internal nature of USDA changed but its relationship with its former external partners in agriculture had also shifted from a tension-filled but generally complementary relationship to one which in specific policy areas were conflict-ridden and often inherently competitive.

In the midst of his reorganization of the department in 1938, Secretary Wallace moved the BAE, then under the leadership of Howard Tolley, to the Office of the Secretary as USDA's policy planning and coordinating arm—that is, a staff agency to the secretary doing research and policy analysis, but first and foremost helping the secretary plan and manage the action of USDA agencies on this newly created battlefield (see Figure 3.1A). By 1942, in the combat that ensued, the BAE had slowly lost significant parts of its planning and coordination role. Wallace was succeeded in 1940 by Secretary Wickard (1940-1945), who was "largely unable to use any kind of policy-making machinery" (Black, 1947:1033). Finally, in 1946, a badly battered BAE was removed from its role as

A. Bureau of Agricultural Economics, 1938-1946

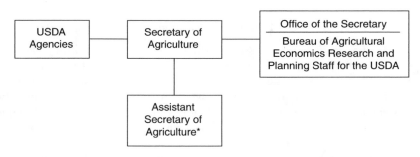

There was only one assistant secretary in USDA. All USDA agencies reported directly to the secretary of agriculture.

B. Bureau of Agricultural Economics, 1946-1953 (Dismantled in 1953)

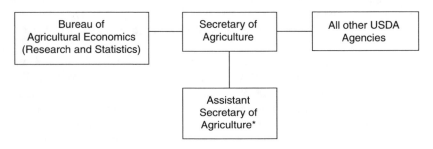

There was only one assistant secretary in USDA. All USDA agencies reported directly to the secretary of agriculture.

FIGURE 3.1 Reporting line for the Bureau of Agricultural Economics (BAE): (A) 1938-1946 and (B) 1946-1953 (dismantled in 1953).

"staff to the secretary" by Clinton Anderson, a former member of Congress and now the new secretary (1945-1948). Anderson returned the BAE to its original role as an economic intelligence and research agency (see Figure 3.1B) (Baker and Rasmussen, 1975:60-62). Paul Appleby, Wallace's deputy and organizational guru, and Harvard professor John D. Black had both argued against the direct involvement of the BAE in policy decisions and anticipated the political dangers for a research unit in such a role (Kirkendall, 1982:232).

Demise and Fragmentation

By the early 1940s, the BAE had acquired a full choir of critics. These critics saw the world in simpler, more monochromatic terms than did economists. In the heat of battle, many critics had great difficulty accepting such elemental economic notions as "opportunity costs" and the "external effects" of policy actions, to say nothing of their ideological difficulty with and rejection of the centralized economic planning effort of the New Deal period in which many economists and the BAE had been so visible. The economic interests in agriculture were by now well organized and the new USDA programs had created a high-stakes struggle for political control.

Critics and enemies of the BAE included many USDA action agencies, which, with congressional support, were determined to do their own planning and resisted any coordination by others—often including the secretary. Other major critics were farmer organizations, especially the most politically influential American Farm Bureau Federation, and the major farm commodity organizations in cotton, wheat, feed grains, etc. In addition, many members of Congress were critical of the BAE, especially conservatives of both parties with important rural and business constituencies who were the major beneficiaries of USDA programs, on which many were now quite dependent (Baker and Rasmussen, 1975:61-64; Hardin, 1946, 1952).

Consequently, following a change in party control from the Truman to the Eisenhower administration in 1953, the new secretary of agriculture, Ezra Taft Benson (1953-1961), a strong critic of government planning and the market interventions of USDA over the previous two decades, announced the abolition of the BAE. Its functions were divided between two new agencies. Farm price, income, and marketing research and data collection went to the Agricultural Marketing Service. Farm management and other economic research went to the Agricultural Research Service (Baker and Rasmussen, 1975:64-65; Wells et al., 1954:1-21). Under the title "The Fragmentation of the BAE," this event was described and discussed by a distressed agricultural economics profession in its *Journal of Farm Economics*. Secretary Benson's case for breaking up the BAE was made by the last chief of the BAE, Oris V. Wells (1946-1953), while all other leaders of the profession in varying degrees were quite critical of the decision and its impact (Wells et al., 1954).

Strong conclusions from this experience were drawn at that time by two of the most distinguished leaders of the profession. Theodore W. Schultz (1954:19) of the University of Chicago (and 1979 Nobel laureate) asked himself "Why have these things happened?" and answered:

Essentially because agricultural economics has been so highly vulnerable to changes in the constellation of political forces within the Executive, the Congress, Farm Bureau and other interest groups and within the far-flung action agencies of the USDA.

To understand the vulnerability of the BAE, one has to appreciate the profound unfriendliness which these organized political forces, both inside and outside government, can feel for agricultural economics research that does not provide the "right" answers . . . .

The powerful AAA of the late thirties was often unfriendly to agricultural economics research, even to the *Agricultural Outlook*. Where an economic analysis touched them, it usually came under AAA fire. . . . The Soil Conservation Service reacted much the same . . . .

. . . To those who were in opposition to the forces represented by the USDA, this BAE effort in . . . planning was simply a Trojan horse to be destroyed, as was soon the case.

The *Agricultural Outlook* referred to above is the BAE's early forward-looking intelligence and market forecasting effort to improve the accuracy of market expectations and pricing efficiency in agricultural markets. The AAA (Agricultural Adjustment Administration) was the action agency responsible for farm price support and production control programs.

Schultz concluded that the economic research function could not be the agent and advocate of major programs. To do so "and thus to maximize its vulnerability, either destroys its objectivity or forces those economists who stay to do 'harmless, descriptive work'" (p. 20). He then listed five guiding principles for achieving orderly and unbiased agricultural economics research in the USDA (pp. 20-21):

1. Agricultural economics research should be placed in a relatively sheltered position in relation to the political instability inherent in the USDA.

2. Agricultural economics research should be so organized that it is relatively independent from (1) the day-to-day staff work of the Secretary's office, (2) the constant, routine, "trouble shooting" work, or the quick program analysis work required by the several action agencies.

3. Agricultural economics research should represent an effort at long-run analysis where competent workers seek to determine the more basic economic characteristics of agriculture and to explain the behavior of these attributes of the economy.

4. Agricultural economics research should be organized to take advantage of the strong complementarity between and among production economics, distribution economics (of which marketing is a part) and of price and income economics. Some important functions also indicate complementarity, for example, The Agricultural Outlook and the publication of Agricultural Economics Research, the preparation of Situation Reports and others.

5. Agricultural economics should be so organized that it has the capacity to recruit and select competent economists and to induce such individuals to join the staff and to make a career as agricultural economists in the USDA.

Professor John D. Black of Harvard, in an earlier 1947 article, reviewed and assessed the BAE experience from 1928 to 1946. His conclusion is similar to that of Professor Schultz (Black, 1947:1035-1036):

> The compelling and sufficient reason why the Bureau should not be the general staff of the Department is *that it cannot safely mix this function with that of collecting and analyzing the data of current change, and with other research which it very much needs to do.* We must therefore have a Bureau solely for these latter functions. The general staff should be attached directly to the Secretary.

> Does this mean that the Bureau should be isolated from the Secretary's Office and from policy making? By no means. It should follow policy and program matters as closely as possible. It should assemble all pertinent data and information bearing on policy and analyze them as closely as possible. It should assemble all possible economic data as to how the different programs of the Department are working out, and weigh and evaluate them. It should even go so far as to predict in detail how alternative program proposals will work out. *But it should not undertake to choose policy, nor even to say what will be the best policy.* This latter is the function of the Secretary (italics in original).

It had begun to be clear early that the provision of objective research and analysis, and even the reporting of market statistics, could not long survive as a USDA function, if not supported internally and defended against external political assault by the administration's political leadership. Assaults came from any organized interest then unhappy with BAE analysis that did not support its program or policy position. The administration was likewise pleased when BAE research supported current programs, but not when objective research failed to support them. Assault sometimes came after a change in administrations, when it was thought that economists had given too much support to the previous administration. In his examination of the "Age of Roosevelt," the historian Richard Kirkendall concluded (1982:257):

> The degree of success that the social scientists enjoyed, however, depended heavily upon their relations with these political men. The years of accomplishment from 1933 to 1940 were years of substantial support from the President and the Secretary of Agriculture; the years of frustration after 1940 were marked by little support for the social scientists from these political leaders.

This support, of course, requires political leaders to have an understanding of how to use economic policy analysis, which, as John D. Black observed, was seriously lacking in some secretaries of agriculture (Black, 1947:1033).

New Deal efforts to cope with the collapse of industrial and agricultural markets during the Great Depression place the accomplishments of the BAE in perspective. The government intervened to stabilize and manage these markets in the National Recovery Act of 1933 (NRA) for industry and the Agricultural Adjustment Act of 1933 and 1938 (AAA) for agriculture. The NRA was soon abandoned as a failure in a conflicted implementation of inconsistent goals, whereas for better or worse, the AAA set the framework for U.S. agricultural policy from the Great Depression into the 1990s. The explanation for the failure of the NRA and the survival of the AAA has been found to lie in the prior development of the agricultural economics profession (beginning around 1900) and its intellectual investment in understanding the problems of agriculture. This led to the creation of the BAE and the presence of a large cadre of economists and other professionals in USDA who did research to understand the nature of agriculture, its problems, and policy issues. Then in the late 1930s, the BAE was to its peril drafted by the secretary to help implement the AAA (Finegold and Skocpol, 1995:Chapters 3 and 4).

In the case of the NRA, there was no prior intellectual investment of comparable scale in the problems and economic behavior of industry. Thus, lacking a cadre of experienced industrial economists in government, the conceptual development and implementation was primarily dependent on politics and recruits from business, most of whom were unable to rise above the experience and interests of their specific firm or industry to address the larger national interest in recovering from the collapse of the U.S. economy. Without the presence of the BAE, it is likely that the AAA would also have failed before it was hardly under way. This investment in economic research and related analytic capacity—and the lack of such in the case of the NRA—is testimony to the value and the need for such capacity in support of public policy (Finegold and Skocpol, 1995).

Many economists today believe the AAA was an unqualified economic policy disaster. Although there have been substantial undesirable economic costs, they forget or ignore the fact that the growing violence and social disorder of the Great Depression years threatened great political costs, including the unraveling of the democratic fabric of U.S. society. The political judgment of the day was that the AAA constituted the least bad, politically feasible option. The BAE favored a direct income subsidy, but its budget costs were high and political beliefs during the depression made direct income transfers philosophically unacceptable. The price supports and production controls that came out of Congress, some variation of which the nation has lived with ever since, are part of the price paid for the smothering of the radical, if not revolutionary, even demagogic ideas then emerging in national and agrarian political discourse. If there was an error, it was in naively believing that the AAA price and production controls were a temporary expedient.

ERS AND LESSONS LEARNED

After eight years of fragmented existence as parts of several divisions in two different line agencies, most of the functions of the old BAE were reassembled but reconfigured in 1961. At the beginning of the Kennedy administration, USDA's economic analysis and research activities were combined as the Economic Research Service. The collection, processing, and dissemination of agricultural statistics for USDA became the responsibility of another free-standing agency, the Statistical Reporting Service (Bowers, 1990; Cochrane, 1961). President Kennedy and his new secretary of agriculture, Orville Freeman (1961-1969), had been persuaded by two presidential campaign advisers, agricultural economists John Kenneth Galbraith and Willard W. Cochrane, to reassemble the functions of the BAE. One of the secretary's first acts was the creation of ERS and the Statistical Reporting Service (later renamed the National Agricultural Statistics Service, or NASS). Subsequently Cochrane was appointed economic adviser to the secretary and director of agricultural economics (at the assistant secretary level). In that role, he presided over the establishment and initial direction of ERS and NASS. Both agencies reported to Cochrane, as did a new small policy advisory unit, the Staff Economists Group, that worked in direct support of Cochrane in his role as economic adviser to the secretary (see Figure 3.2) (Cochrane, 1961, 1965).

The mission envisioned for ERS in its early days is an important baseline for understanding ERS today and the changes that have taken place as it approaches the fourth decade of its existence. Cochrane described that mission clearly. He made the case that ERS should be viewed as a "staff agency to the Nation" (Cochrane, 1983:30):

> It must be prepared to respond regularly and effectively, without compromising itself, to the economic analytical needs of the Office of the Secretary; it must understand and appreciate the intelligence needs of members of the Congress and find ways of satisfying those needs without coming into conflict with the administration in power; and it must recognize and anticipate the information and intelligence needs of a diverse national public and develop effective channels for meeting those diverse needs.

Both Cochrane and his successor as director of agricultural economics, John Schnittker, believed that ERS should be responsible for all agricultural economics work in the Department of Agriculture.

The ERS has been in existence for 37 years, only slightly longer than the life span of the BAE. Has the mission of ERS changed in any fundamental way since the 1960s? If so, how? What lessons can be learned from its experience? Meaningful evaluation of ERS's research program and its performance requires a clear answer.

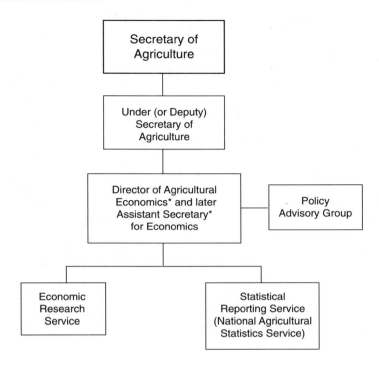

* All have been senior agricultural economists.

FIGURE 3.2 Reporting line for the Economic Research Service (ERS), 1961-1993.

Policy and Political Environment

Although, over the history of the BAE, the bulk of its resources was devoted to the problems of agricultural production, marketing issues were added to its agenda well before its demise in 1953. Marketing research expanded further during 1953-1961, the period when economic research in marketing was part of the Agricultural Marketing Service. Over the same period, increased attention was also being given to work on "hired farm workers, local studies concerning old age, health services, levels of living and rural industries" by former BAE economists now in the Agricultural Research Service (Koffsky, 1966:415).

The BAE and ERS faced rather different working environments. One clear difference is found in the structure of their reporting line (Bowers, 1990:234):

> The differences between ERS and the BAE were obvious. The new agency reported not to the Secretary, as the Bureau did, but to the Director of Agricultural Economics. It had no planning functions and political work was carefully

isolated in the Director's office in order to preserve the objectivity of ERS work. ERS Administrators would be civil service employees [and] the political side of staff work would be handled by a Staff Economists Group.

The BAE, along with the Bureau of Labor (which later became the Bureau of Labor Statistics), were the earliest experiments in developing systematic economic research to support federal policy (Duncan and Shelton, 1978:6-7). This BAE experiment occurred in the midst of an even greater experiment in government action to deal with the threatened breakdown of the fabric of the society and economy during the Great Depression and, without a pause, the demands of World War II. Thus, the BAE may be said to have worked in revolutionary, certainly politically turbulent, times. By 1961, when ERS was founded, policy for agriculture had settled into somewhat more stable patterns. Change, even major change, continued to occur, but it was by comparison evolutionary. Many of the same forces buffeted ERS, but they were less dramatic in scale, thus less visible and less inclined to involve highly visible political duels to the death. As a result, this review of ERS history focuses not on a few dramatic events so much as on the often-unexciting details of evolving changes in (1) the capacities of ERS, (2) agriculture and society that modify policy and political agendas, (3) ERS research and intelligence produced in response to policy need, and (4) ERS internal and external organization made in support of the evolving agenda of ERS research and information products.

It is also the case that the BAE operated in a relatively closed national economic setting, in which international trade, monetary, and fiscal impacts had little influence on the U.S. food and agriculture system. Beginning in the 1960s, this changed, as a large international capital market began to emerge and to expand rapidly through the 1970s and 1980s. Thus, early in its existence ERS was confronted with a U.S. agricultural and food economy in which global production, trade, capital market movements, and other international macroeconomic forces and policies started to have major impacts on the performance and problems of the U.S. food and agriculture system. Consequently, the research and policy analysis problems facing ERS grew progressively more complex. Indeed, the development of secondary data and analysis, such as the various indicators and the *Situation and Outlook* data and analysis also became more of a challenge. Understanding and explaining sector performance in an open economy is inherently more difficult.

As this external environment, including political issues and policy decisions facing USDA, has changed, so have the opportunities and imperatives driving ERS information, policy analysis, and research output. Major changes in the food and agriculture sectors, changes in society's need for research, analysis, and information products, and a new administration's political and policy agenda are often reflected in internal changes in the organization of ERS.

The 1960s: A New Beginning in a Different Era

The ERS faced a rather different and evolving pattern of policy issues and research needs in the 1960s, compared with the issues that faced the BAE of the 1930s and the 1940s. One such change involved the function of foreign market development, intelligence, and research. The interest in foreign markets was first given division status in 1930, as the Foreign Agricultural Service Division of the BAE. However, this division was transferred out of the BAE to the Office of the Secretary in 1938 as its action arm for export promotion to reduce farm surpluses. The growing support for export promotion during the depression was driven by the popular political belief that one could "export the farm problem" by exporting the surpluses that were depressing prices and incomes. Except for the period of World War II, agricultural exports were limited in this era. The small Foreign Agricultural Analysis Division and some functions of trade policy were transferred back from the secretary's office to the newly created ERS when the economic research, analysis, and secondary data development functions of USDA were reassembled in 1961 (Baker et al., 1963:498-500, 505). The initial organization of ERS included three domestic and two foreign economic intelligence, policy analysis, and research divisions: (1) Economic and Statistical Analysis, (2) Farm Economics, (3) Marketing Economics, (4) Development and Trade Analysis, and (5) Regional Analysis (Box 3.1; Cochrane, 1961:73).

The 1960s saw the beginnings of major postwar domestic political changes. Reapportionment of the House of Representatives in 1960 and 1970 and the Supreme Court's "one-man, one-vote" decision in 1964 increased urban, consumer, and labor influence while decreasing rural and farm influence. Enactment of the 1965 farm bill saw the first evidence in agricultural politics of an organized labor, welfare, and consumer coalition with interests in low food prices and food stamps for striking workers and the poor. The agricultural committees in Congress were forced to combine all their previously separate commodity support and other agricultural bills into one large "omnibus" legislative package. Agricultural interests, despite their differences, had to stand together in a common coalition in order to achieve their major legislative goals. This omnibus legislation now encompasses not only traditional farm programs but also policy for resource conservation and the environment, P.L. 480 subsidized and free foreign food aid, domestic food assistance, rural development, research, and extension. By the end of the 1960s, this developing coalition of diverse interests was necessary to achieve a majority vote for passage of farm legislation—as well as for the other pieces of the omnibus package (Bonnen, 1980).

President Johnson's Great Society programs significantly expanded the USDA role in assisting the poor with food stamps, school lunches, and improved nutrition, although, in the last half of the 1960s, budget pressures of the Vietnam conflict reduced or constrained expenditures on domestic programs. This period is the origin of increased concern about food safety and environmental quality.

BOX 3.1
Division Organization of the Economic Research Service, 1961-1998

1961
- Development and Trade Analysis
- Economic and Statistical Analysis
- Farm Economics
- Marketing Economics
- Regional Analysis

1962
Farm Economics was split into two divisions:
- *Farm Production Economics*
- *Resource Development Economics*

1965
Resource Development Economics was split into two divisions:
- *Economic Development*
- *Natural Resource Economics*

1965 Organization
- Economic Development
- Economic and Statistical Analysis
- Farm Production Economics
- Foreign Development and Trade
- Foreign Regional Analysis
- Marketing Economics
- Natural Resource Economics

1971
Foreign Economic Development Service transferred to ERS from the secretary's office as the Foreign Development Division

1973
Three divisions were dissolved:
- *Economic and Statistical Analysis*
- *Farm Production Economics*
- *Marketing Economics*

Two new divisions were created:
- *Commodity Economics*
- *National Economic Analysis*

Five other existing divisions continued:
- Community and Human Resources
- Foreign Demand and Competition
- Foreign Development
- Natural Resource Economics
- Resource and Development Economics

1977
The *Economics, Statistics and Cooperatives Service* (ESCS) was created by combining ERS, SRS, and FCS as separate units under one administrator.

ERS divisions remained much the same as in 1973:
- Commodity Economics
- Economic Development (from earlier merger of Community and Human Resources Division and the Natural Resource Economics Division)

- Foreign Demand and Competition
- National Economic Analysis

Foreign Development transferred from ERS to newly created Office of International Cooperation and Development

Reorganization September 1981
ESCS disbanded, returning ERS and SRS to agency status; ERS division organization had been changed during ESCS period to:
- Economic Development
- International Economics
- National Economics (merger of Commodity Economics and National Economic Analysis divisions)
- Natural Resource Economics (pulled back out of Economic Development)

1983
ERS field staff disbanded but division organization remained unchanged

1987
- Commodity Economics
- Agriculture and Trade Analysis
- Agriculture and Rural Economic Resources and Technology

October 1994 Reorganization
- Commercial Agriculture
- Food and Consumer Economics
- Natural Resources and Environment
- Rural Economy

October 1997 Reorganization
- Food and Rural Economics
- Market and Trade Economics
- Resource Economics

Sources: Baker and Rasmussen (1975), Bowers (1990), Cochrane (1961), Koffsky (1966) and materials supplied by the Economic Research Service, 1998b.

There was also a growing policy focus on problems of low-income rural people and the communities "left behind" by the industrial development of agriculture (Daft, 1991:149). Thus began another fundamental but slow transformation of the Department of Agriculture from its near exclusive focus on farmers and agricultural markets eventually to include, as also important, public concern over nonfarm rural people, depressed rural regions, environmental quality, domestic and imported farm labor, and the consumer's interest in food safety, adequate nutrition, and access of low-income consumers to food. These concerns began to change the research agenda of ERS. The great expansion in action and regulatory budgets in these new areas, however, did not begin to approach the scale of the USDA farm program budgets until the early 1970s.

Attempts to ensure adequate nutrition for the poor began during the period of massive unemployment in the 1930s. Fred Waugh, a BAE and later an ERS economist, developed the economic concept of food stamps and did the analysis that established its feasibility as a more effective alternative to the direct physical distribution of food. After a limited experiment in 1939-1943 and a larger one in 1961-1964, a permanent, nationwide food stamp program was enacted in 1964 (Tweeten, 1979:387-395).

During the early 1960s, ERS became deeply involved in research on foreign trade. The U.S. policy emphasis on P.L. 480 emergency food aid and subsidized farm exports for market development abroad created pressures and opportunities for ERS to expand its research in this area (Baker and Rasmussen, 1975:66). Also, the Kennedy round (1963-1967) of negotiations under the General Agreement on Tariffs and Trade (GATT) began an international effort to address agricultural trade restrictions. This focus continued through the Tokyo round (1973-1979) and the most recent or Uruguay round (1986-1994) of trade negotiations (Tweeten, 1992:211-215).

In addition to foreign trade research, over the first half of the 1960s, ERS responded to new policy initiatives of the Kennedy-Johnson administration with expanded efforts in natural resource policy issues, rural economic development, river basin and watershed development, labor and employment issues, and low-income and poverty problems in rural America (Bowers, 1990:238).

Although over 70 percent of the first ERS budget went to traditional farm economics and marketing work, the initial division-level organization in 1961 had already begun to reflect the new policy agenda in foreign trade and development (Cochrane, 1961:72-73; Bowers, 1990:237). Then in December 1962 the Resource Development Economics Division was established to focus on rural resource issues and rural development. By 1965, this new division had been divided into the Economic Development and the Natural Resource Economics divisions. At the same time, the Development and Trade Analysis Division was renamed the Foreign Development and Trade Division to differentiate it from the domestic focus of the new Economic Development Division (Cochrane 1983:31; Baker and Rasmussen, 1975:65-67; Bowers, 1990:238). By 1965, less than half of all ERS publications from in-house research dealt with the conventional agricultural subject matters of the BAE period (Koffsky, 1966:415). The research, analysis, and secondary data development agenda of ERS was already evolving (Box 3.1).

Willard Cochrane resigned as director of agricultural economics in 1964 and was replaced by the director of his policy staff, John Schnittker. The first administrator of ERS, Nathan Koffsky (1961-1965), became director of agricultural economics in 1965 when Schnittker left and was succeeded as ERS administrator by M.L. Upchurch (1965-1972). Over the decade of their collective tenure, ERS appropriations and personnel numbers rose and then fell back to their original fiscal 1962 levels. The expanding number of research topics created stress on ERS resources that could be alleviated only partially by reallocation and reorganization. Budget transfers from other agencies, such as the Soil Conservation Service and the Agency for International Development, made the expansion in the scope of ERS research possible, especially in the last half of the 1960s (Bowers, 1990:237; Baker and Rasmussen, 1975:67-68).

Koffsky and Upchurch both struggled with the problem of finding the resources for and sustaining the appropriate balance between ERS-initiated long-term or basic lines of research inquiry and the externally mandated or requested problem analysis. ERS found that they had political support or demand for short-run provision of secondary data development and analysis relevant to current economic problems and political perceptions, but there was little recognition or support for investing in the longer-term foundation of knowledge necessary to sustain high-quality analytic work, whether short-term or long-term (Baker and Rasmussen, 1975:67-68). What is most commonly not understood is that, especially when developing new problem areas for analytic work, basic research is needed to create the concepts, measurement techniques, and models to address applied problems (Johnson, 1986:Chapter 2). This is required to identify and understand the important characteristics, changing structure, and behaviors of the problem. In contrast, the need for and acceptance of "quick and dirty" policy information are inherent needs that result from the time pressures faced by political-level decision makers. With limited or declining resources, the necessity to service large numbers of immediate decisions can make it difficult for a research and analysis unit to maintain the high quality of analytic product that its reputation and support depend on in the long run.

The 1970s: A Tumultuous Decade

A major economic shock in 1972-1973 shook the world. The United States shifted over the 1971-1973 period from a fixed to a flexible exchange rate regime, effectively devaluing the dollar and also leaving U.S. domestic markets more open to the effects of international market events. Widespread drought led to a series of national crop failures. The Soviet Union experienced extensive crop failures and, without warning and for the first time, contracted to import large amounts of grain instead of rationing their short crop. The world was left with a major shortage of grain and an unprecedented increase in food prices (Hathaway, 1974; Schuh, 1974). This shock was compounded for consumers by the inflationary effect of the newly formed Organization of Petroleum Exporting Countries (OPEC) cartel's reduction in oil production, causing a sharp rise in energy prices and a flood of money from OPEC country profits into the world's financial system. World market commodity prices, especially primary products, rose with inflation (Tweeten, 1989:336-341; Cochrane, 1993:150-170).

In the face of wide fluctuations in domestic food prices, the United States imposed a series of export embargoes and constraints on surplus disposal in a mostly symbolic effort to protect and stabilize some domestic prices. The embargoes had limited impact and created more problems than they solved, but the situation generated increased demands on ERS domestic and international agricultural research and market analysis (Economic Research Service, 1986).

Over the same period, the United States was involved in the Tokyo round of the GATT trade negotiations (1973-1979) and also was experiencing a massive growth in agricultural exports. Between 1970 and 1980, U.S. agricultural exports grew from $7.3 to $41.2 billion—effectively internationalizing the U.S. agricultural sector (Council of Economic Advisers, 1998:397).

This export boom was made possible by several events. As mentioned above, an overvalued dollar was devalued. Sustained world economic growth, especially in the developing countries, increased world food demand. And agricultural legislation from the 1960s to 1973 changed U.S. agricultural policy from a regime of high price supports, set well above world prices, to low prices with a compensatory direct payment from the U.S. Treasury to farmers to make up the difference in farm income. This policy change shifted much of the cost of farm income support from consumers to taxpayers and began to give U.S. farmers sustained access to world markets for the first time since price supports and acreage controls were introduced in the 1930s. Farm prices became far more volatile, creating new challenges for ERS price forecasters.

Interest in rural development and the failing vitality of many rural communities grew in Congress during the 1970s, but this interest was not shared by the Nixon administration (Daft, 1991:149). It had by then become clear how fallacious was the traditional assumption that the farm programs constituted a policy sufficient to sustain all of rural America. ERS had begun to construct the first reasonably systematic data base for rural development. Although the United States has a large data base on the urban population concentrations of metropolitan areas, its rural areas of low population density are more costly per 1,000 of the population surveyed and continue to lack adequate and reliable data for most policy purposes (National Research Council, 1981).

Until the 1970s, the federal role in national environmental regulation was focused on managing public lands, soil conservation, and legislative responses to specific issues. The passage of the National Environmental Policy Act of 1969 began to move the country toward a broader national program of environmental protection and management. By the end of the 1970s, the United States was enacting and enforcing national regulations across a broad array of national concerns: water quality, air quality, pesticides and toxic chemicals, hazardous wastes, occupational health and safety, mine safety, ocean pollution, coastal zone protection, and resource conservation and recovery (Council on Environmental Quality, 1979-1984).

The context of environmental and resource conservation policy in agriculture also began to change in the 1970s. The soil conservation programs, begun in the drought conditions of the Great Depression, were seen as an integral part of the farm programs and helped create support for the production controls and price supports. From the 1930s to the 1970s, soil conservation goals and production adjustment objectives were viewed as mutually supportive of each other. The worldwide growth in effective demand for food and the consequent export boom and high prices of the 1970s led to elimination of U.S. production controls in 1974 through 1977 and to intensification of farm production and a major change in land use. Grasslands were plowed up for crop production; highly erodible land was brought back into production; and terraces, windbreaks, and shelterbelts that took years to establish were torn out. The conservation practices encouraged by government policy since the Great Depression were swept away (Potter, 1998:19-26). It was clear from ERS research that the commodity programs as administered had become destructive of and no longer consistent with any commitment to soil conservation (Reichelderfer, 1985).

A broader environmental consciousness of human impacts on the balance of nature had begun 15 years earlier with Rachel Carson's *Silent Spring* (1962). By the end of the 1970s and the early 1980s, environmental and natural resource protection advocates and organized interests had begun to be major players in the public debate over USDA legislation. All of this influenced discussion of agricultural and rural resource uses in major ways and began a steady growth in USDA budgets for environmental and natural resource research and regulation (Figure 3.3). The capacity of ERS to deal with this diverse set of natural resource issues built on and was sustained in part by its long investment in the data base on and analysis of land use. This work began in a significant way in the Division of Land Economics in the 1920s under L.C. Gray. In the 1970s, land use work for policy purposes ranged from major river basin planning efforts with the Soil Conservation Service, the Forest Service, and others, evaluation of major watershed projects, foreign ownership of U.S. farmland, soil and water conservation, resource inventories, farmland protection, water quality assessment, and the impact of technologies on land and studies of ground water contamination (Cotner and Heneberry, 1991).

The politics of agriculture continued to change over the 1970s. The intense specialization and commercialization of production that accompanied rapid increases in agricultural productivity since the 1950s had, by the decade of the 1970s, fragmented the economic interests in agriculture. Conflict between farm producer interests and various other elements of the food system made the legislative coalition in agriculture and the food system increasingly contentious and unstable. The White House and the secretary of agriculture were now commonly confronted with no-win situations in which there was little political incentive to provide legislative leadership. Yet they often still had to settle policy issues that were forced to the secretary and White House levels by otherwise unresolved conflict. By the early 1970s, passage of agricultural legislation was not possible without at least a tacit coalition of consumer, labor, and urban welfare interests. Food stamps had become the glue that held sufficient political support in place to

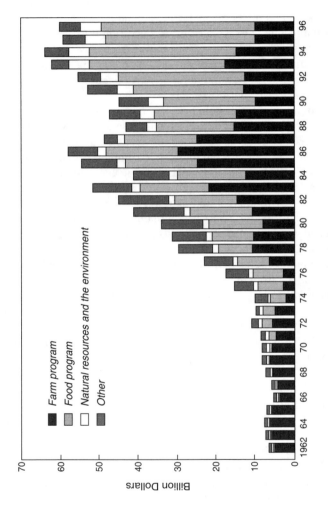

FIGURE 3.3 USDA outlays by major program category, 1962-1996. Source: Economic Research Service, Historical Budget Outlays, electronic data base, U.S. Department of Agriculture, March 1995.

pass "farm legislation" (Bonnen, 1980). The largest impact on USDA budgets during the 1970s came from rapid expansion in the food stamp program during the Nixon-Ford and Carter administrations (Figure 3.3).

The main consequence for ERS of the tumultuous decade of the 1970s was an agenda for research and analysis that expanded to cover a more diverse range of topics than it had addressed in the 1960s. It dealt with not only the effects of the major events of the early 1970s on domestic and foreign agricultural markets of the United States, but also the impacts of the subsequent U.S. export embargoes on those markets of 1973-1975 and 1980 (Economic Research Service, 1986). The Food and Agriculture Act of 1977 shifted farm income support calculations from a parity price basis to a more complex and contestable cost of production measure, which the law then specifically required ERS to produce. ERS has subsequently adapted these cost of production indicators for individual crops to the many subsequent changes in farm legislation without getting embroiled in abrasive political disputes.

The commodity market-oriented *Situation and Outlook* (*S&O*) program (described in Chapter 5), begun in the 1920s, reached a highly sophisticated analytic capacity in the 1970s. ERS increasingly developed a commodity-oriented organizational framework, which integrated all related intermediate and long-term research (including work on econometric methods), policy analysis, market intelligence, and the supporting secondary data and information base. ERS pioneered in the development of econometric models of national and later international agricultural commodity markets now commonly in use. An aggregate U.S. agricultural sector model was developed for assessing commodity and regional sector interactions and other policy impacts. This approach brought together commodity teams that included econometricians, market analysts (with institutional knowledge and informational networks across the market), policy analysts (with policy informational networks), and experienced statistical clerks (who could quickly spot data aberrations). The *S&O* teams interacted almost continuously in a process that both tracked and simulated the agricultural sector and its policy and market changes. The development of this *S&O* system capacity is described by Abner Womack (1991).

During the 1970s, ERS provided important analysis of the impact on U.S. markets and on its trading partners of the fundamental shift in U.S. farm policy from direct price controls to letting markets set prices. It provided research in support of the U.S. Tokyo round of trade negotiations (1973 to 1979) under GATT. Work was done on the potential for sending U.S. agricultural exports to selected developing country markets and on some of the related economic development problems of the low-income nations of the world. ERS began conducting research on the rapidly changing structure and productivity of the U.S. agricultural sector. A growing range of environmental issues and regulations was addressed and natural resource economics problems explored. In assessing this challenge, the USDA historian, Douglas Bowers observed (1990:236-237):

"without increases in personnel, the only way to undertake new research was to cut back somewhere else. . . . The ability of ERS to shift into new areas became key to its success." One area of ERS to feel significant budget erosion in the 1970s was rural development.

This growing scope of research and analysis was done in a context in which ERS personnel numbers fluctuated around an average of 1,000—of which about half or less were economists. An inflation-adjusted ERS budget began the decade in the low $60 millions, declined a bit by mid-decade, and by 1970 had risen to just over $70 million in constant 1996 dollars (Tables 3.1 and 3.2). Quentin West, who succeeded Upchurch as ERS administrator in 1972, was somewhat more successful than his predecessors with congressional appropriations for ERS. It has also been observed, in reference to the 1960s and 1970s, that "in the long run, it was probably the continued backing for ERS from the Secretary's Office that was the most important factor in winning more support from Congress" (Bowers, 1990:236).

West soon presided over another reorganization of ERS division structure while experimenting with a matrix management system of temporary task groups to address specific problems as they arose, rather than assigning such work to a division. The experiment pushed ERS further toward more short-term service work and was often disruptive of other ERS organization and research activities. The experiment ended when West left for another USDA position in 1977 and was succeeded by Kenneth Farrell (Rasmussen, 1991:87).

Another experiment began in 1977, and much energy was spent in the creation, internal resistance to, and management of an "ill-considered and unworkable" administrative structure that was the result of a USDA response to a Carter administration commitment to reduce the number of government agencies (Rasmussen, 1991:87-88). This cosmetic effort removed ERS, the Statistical Reporting Service, and the Farmer Cooperatives Service from agency status and placed them under the direction of the administrator of a new agency, the Economics, Statistics and Cooperatives Service (ESCS). This arrangement did little more than add an additional administrative layer to the reporting line. The incoming Reagan administration disbanded ESCS in September 1981, returning ERS and SRS to agency status (Rasmussen, 1991:87-88). The pressure from its clientele had extracted the Farmer Cooperatives Service from ESCS a bit earlier.

The 1980s: Farm Crisis and Recession

The decade of the 1980s saw an end to the economic expansion of the 1970s. Farmers had faced strong incentives in the 1970s to expand production and to invest in new productive capacity. Acreage controls were dropped during the 1974-1977 period, when expanding effective world demand for food caused market prices for U.S. farm products to run well above government support levels. In addition, negative real interest rates prevailed over much of the decade of the

TABLE 3.1 Funds Appropriated for the Economic Research Service (ERS), Fiscal 1962 to Fiscal 1996

Fiscal Year	Appropriated Funds ($ million)	Inflation Index (1987=100)	Appropriations in Millions of 1996 Dollars	Transfers to ERS in Millions of 1996 Dollars	Total Funding in Millions of 1996 Dollars
1962	9.7	22.6	58.6	7.9	66.5
1963	9.5	23.1	56.2	10.1	66.2
1964	9.2	23.7	53.0	20.2	73.2
1965	10.9	24.4	61.0	21.3	82.3
1966	11.8	25	64.5	25.7	90.2
1967	12.4	26.1	64.9	27.2	92.1
1968	12.8	27.8	62.9	26.0	88.9
1969	13.4	29.2	62.7	24.3	87.0
1970	14.6	31.6	63.1	16.4	79.5
1971	16.1	34	64.7	14.5	79.1
1972	16.5	39.3	57.4	13.6	70.9
1973	18.1	41.9	59.0	11.4	70.4
1974	19.6	45.5	58.8	10.2	69.1
1975	22.4	51.2	59.8	10.4	70.2
1976	25.8	54.1	65.1	10.4	75.5
1977	28.0	57.7	66.3	9.5	75.8
1978	31.7	61.7	70.2	11.3	81.5
1979	35.2	66.4	72.4	11.1	83.5
1980	36.5	73.3	68.0	11.2	79.2
1981	39.5	82.1	65.7	10.3	76.0
1982	39.4	85.9	62.7	8.0	70.6
1983	39.0	89.5	59.5	8.4	67.9
1984	44.3	91.3	66.3	4.0	70.3
1985	46.6	95.7	66.5	3.6	70.1
1986	44.0	98.6	61.0	3.3	64.3
1987	45.0	100	61.5	2.6	64.1
1988	48.2	101.4	64.9	2.0	67.0
1989	49.3	107.5	62.6	2.5	65.2
1990	50.7	111.5	62.1	2.5	64.6
1991	54.4	116.5	63.8	3.4	67.2
1992	58.7	119.8	66.9	7.2	74.1
1993	58.7	124.3	64.5	11.5	76.0
1994	55.2	129.1	58.4	9.9	68.4
1995	53.9	(est) 132.7	55.5	10.0	65.5
1996	55.1	(est) 136.6	55.1	7.2	62.3

Source: Information Services Division of the Economic Research Service, U.S. Department of Agriculture. The inflation index is based on federal government purchases (1962-1971) and federal government nondefense purchases (1972-1994) as published by the Bureau of Economic Analysis, U.S. Department of Commerce. Data for 1995-1996 are estimates based on change in the Consumer Price Index.

TABLE 3.2 Economic Research Service Personnel Fiscal 1962-1998 (in full-time equivalents)

Fiscal Year	Total FTEs	Fiscal Year	Total FTEs	Fiscal Year	Total FTEs
1962	1,095	1975	1,057	1988	791
1963	1,000	1976	1,059	1989	792
1964	1,073	1977	945	1990	780
1965	1,159	1978	970	1991	773
1966	1,142	1979	933	1992	804
1967	1,189	1980	1,039	1993	795
1968	1,222	1981	993	1994	718
1969	1,059	1982	946	1995	625
1970	1,072	1983	903	1996	591
1971	1,118	1984	876	1997	570
1972	1,044	1985	866	1998	554
1973	1,030	1986	846		
1974	1,021	1987	813		

Source: Information Services Division of the Economic Research Service, U.S. Department of Agriculture.

1970s, as inflation galloped ahead of interest rates. Market expectations were for worldwide food shortages and a rising real cost of food. With such inducements, many farmers borrowed heavily to increase their productive capacity (Tweeten, 1989: 343-348; Cochrane, 1993: 150-170).

When inflation reached double-digit levels by the late 1970s, the Federal Reserve Board slammed on the brakes with double digit interest rates in October 1979, abruptly ceasing to fund the budget deficit. The financial markets responded. The real value of the dollar, which had been declining in the 1970s, began appreciating rapidly through the first half of the 1980s, making U.S. exports more expensive (Council of Economic Advisers, 1998:247-248). Also, the world economy slid into recession in the early 1980s, cutting economic growth and further reducing the effective demand for U.S. agricultural exports.

The large inflationary flow of OPEC monetary assets into the international financial system in the 1970s created easy access to borrowed capital and a buildup of national debt in developing nations, especially in Latin America. As recession set in, and with rising interest rates in the 1980s, some countries had to refinance or face default on their debt. This financial crisis added to the impact of the recession on world trade. U.S. macroeconomic policy in the 1980s added to the international problem by running a mix of tight monetary policy with an expansionary fiscal policy. The resulting large and growing federal budget deficits had to be financed in part from international capital markets, putting further pressure on interest rates (Council of Economic Advisers, 1998:247-248).

As effective demand for U.S. agricultural products declined in the recession of the 1980s, U.S. farm prices dropped drastically along with farm land values, while input costs rose at an even faster pace. Instead of the expected food shortages, surplus stocks and program costs grew rapidly. This recession and excess agricultural capacity created the 1982-1986 farm crisis in which many farmers who had borrowed most heavily during the expansion of the 1970s faced a negative cash flow and were often unable to pay the carrying cost of their debt. Farm bankruptcies became widespread as farmers faced the most severe financial problems since the Great Depression (Stam et al., 1991). It also turned the USDA and ERS agendas toward a greater focus on farmers' problems.

Unfortunately, the 1981 farm legislation had assumed that the inflation, food shortage, and demand expansion of the 1970s would continue indefinitely. Congress wrote into legislation a specific set of rising price supports that exceeded the declining world market prices in the early 1980s. Although farm income was protected by deficiency payments, U.S. exports lost market share to competitors, and growing government-held Commodity Credit Corporation stocks burdened the market (Tweeten, 1989:341-345; Cochrane, 1993:154).

By 1982, the new administration in the USDA not only faced an entirely different set of problems than they had anticipated in 1981, but also suddenly and unexpectedly they had to reverse policy direction—from a philosophically comfortable free market approach to one of aggressive government management of the market (Lesher, 1991). It remained for the 1985 Farm Security Act to repair a self-inflicted wound that delayed the agricultural sector's recovery (Tweeten, 1989:341-345). U.S. farm exports began to rebound after 1985 as the world recovered from recession, and the 1985 agricultural legislation programmed U.S. price support loan rates to stay well below world market prices. This recovery coincided with the start of the Uruguay round of trade negotiations under GATT (1986-1994), in which a major U.S. objective was reduction of agricultural trade barriers (and thus farm subsidies), especially in the European Common Market.

Through each round of trade negotiations, ERS had become more involved, developed greater capacity, and provided more economic research and analysis on agricultural trade issues. ERS had a significant impact on the Uruguay round (1986 to 1994) through analytic support of the secretary and the U.S. trade negotiators. A major contribution was the provision of a single quantitative measure of the different barriers to trade calculated for the 40 largest trading countries—as we said in Chapter 2, identifying winners and losers in agricultural trade. Its "producer subsidy equivalent" provided a comparable measure of how badly a nation was sinning against the ideal of free trade. A similar measure of consumer food subsidies, a "consumer subsidy equivalent," was also provided. These synthetic measures had a significant influence in framing the issues for negotiations between participating countries, although estimates for some specific commodities did not result in entirely defensible results.

Two unavoidable facts condition the future of U.S. agricultural trade. First, any significant expansion in demand for U.S. farm output now depends on exports. Expansion in domestic food demand is severely limited by the slow growth in the U.S. population combined with almost no increased demand for farm food production as income increases. At the high per capita income levels of an industrial nation, the U.S. already consumes about as much food as physically possible. Thus, economic intelligence and research on agricultural trade will continue to be a major item on the political agenda of farmers, farm input firms, and the food industry. Second, the demand for U.S. food and agricultural exports depends on continued economic growth in low- and middle-income developing nations and improved equity in the distribution of income in those nations.

It was 1994 before farm exports regained their 1981 levels and 1995 before the net worth of U.S. farms regained their 1980 level (Stam et al., 1991). By the late 1980s, farm policy and farm program budgets (and subsidies) were being politically constrained by the pressures to reduce trade barriers in agricultural markets and by efforts to slow a burgeoning U.S. budget deficit. Throughout the 1980s, ERS economic analysts and researchers faced the pressures of a full and an increasingly diverse agenda of demands (Lesher, 1991).

New environmental and food-related interest and advocacy groups were propelling legislative change that affected the ERS agenda. By 1980, the involvement and influence of environmental interest groups in negotiations on farm legislation was unavoidable. By the middle 1980s, passage of farm legislation was not possible without satisfying some of the major goals of environmental interests. The pressure to restrain the rapidly expanding budget deficit was turning federal budget making into a zero-sum game, which made farm bills even more difficult to pass than had been the case in the 1970s. Broader political coalitions were now necessary for enactment of farm legislation. Growing public concern over the environmental effects of farm programs made the more pragmatic of organized environmental interests attractive legislative partners for farm interests. As a consequence, the radical notion of the early 1980s of forcing farmers to comply with conservation requirements to be eligible for commodity program support became law in the Farm Security Act of 1985. This form of cross-compliance provided assurance that farm and environmental policies were consistent, muting environmental criticism of farm programs. Plowing up erodible soils ("sod busting") or draining remaining wetland ecosystems ("swamp busting") now made one ineligible for farm program support. Major policy questions were raised in this change. Indeed, the change itself was fueled in part by the factual base provided by prior environmental research done in ERS (Potter, 1998:50-51, 56-57, 61-64).

Over the decade of the 1980s, USDA publications show that ERS research and analysis covered a still-widening spectrum of topics, but at the same time, due to the farm crisis, it increased the focus on problems related to agriculture in production, finance, trade, environmental pollution, and natural resource sustainability and conservation. ERS prepared background materials and analysis for the 1981 Agriculture and Food Act, the 1983 payment-in-kind legislation, the critical Food Security Act of 1985, and the 1990 Food, Agriculture, Conservation and Trade Act. The increasing complexity of the titles on these farm acts is indicative of the growing diversity of the subject matter and the political coalitions involved in the legislation. This broader agenda meant that, besides agriculture, ERS was also doing work on food stamps, direct food distribution, and food safety, rural community development and welfare problems, and rural-urban land use, and it was putting more effort into environmental and natural resource issues, often in collaboration with other agencies in federal and state government.

Political and policy support for rural community development, which had increased in the late 1970s, began to wane in the 1980s (Daft, 1991:150). Effective policy interest in rural development has long experienced wide, repeated swings between strong interest and support and almost none. In part, the lack of consistent political support appears due to the great diversity of rural interests and a consequent lack of an organized voice or coalition of voices able to sustain rural interests at state, local, and national levels. Despite these fluctuations in support, ERS has continued to develop and to maintain the only systematic, national data base for understanding and analyzing the problems of rural development. A major ERS innovation in the 1980s created greater capacity for generalizing causal relationships in rural development, by developing a standard, quantified classification of the primary categories of economic activity for all U.S. rural counties. Without this sustained ERS data base, much of the comparative rural development work done at the state level and in universities would not be as useful or even possible in some cases.

Even ERS's agriculturally related work grew more diverse in the 1980s. Publications on new topics in agriculture or those demonstrating increased frequency included, for example, pesticides, energy requirements and costs, foreign ownership of farm land, changing structure of farming, economies of farm size, corporate farming, ethanol, bovine somatotropin (a hormone) use in dairy production, and new farm regulations on sod busting and swamp busting, to name a few examples.

ERS has faced difficult problems in maintaining the resource base and organization of its highly effective *S&O* teams. As the organization of ERS shifted to deal with a broadening scope of issues beyond agriculture and as its resource base shrank, the *S&O* teams lost resources and support. Martin Abel (1991) has described several weaknesses that developed by the late 1980s. Without the integrative effects of *S&O* teams, the research, policy analysis, and information and intelligence functions of ERS, although still strong complements, became more difficult for management to coordinate. By the mid-1990s, the ERS investment in *S&O* commodity market analysis had become perilously thin. Compounding the problem was the late 1970s creation of an independent World Agricultural Outlook Board. This change made organizational sense, but it also transferred professionals and funds away from ERS and created ambiguity and uncertainty about the ERS role and responsibilities in the *S&O* programs of the department that have not yet been resolved. This is indicative of a growing organizational issue that USDA and ERS face today in producing and mobilizing information and analysis of policy problems. Policy issues increasingly require collaboration of diverse USDA (and often non-USDA) agencies to create and interpret the necessary information and analysis. In these cases, no single agency, including ERS, can by itself ensure the production of a high-quality, relevant product. Leadership for such an effort must come from the secretary's level, and on occasion from the interagency cabinet level.

As a consequence of the farm crisis, USDA expenditures on farm programs again grew to be the largest USDA budget category through the mid-1980s. Despite their high rate of growth, food stamp and nutrition programs exceeded farm program budgets in only four years in the 1970s and five years in the 1980s. Farm program expenditures fluctuate widely and have been largest when agricultural market prices are low, surplus stocks high, and farmers in financial trouble. Since 1988, however, the still-expanding food and nutrition budget has consistently been the largest expenditure category. By the end of the 1980s, it had grown to 50 percent of the USDA budget. Today, food and nutrition expenditures are annually 60 percent or more of the total USDA budget (Figure 3.3). Food stamps account for most of these expenditures (Figure 3.4). Natural resource and environment programs have also continued to grow and, in the mid- to late 1990s, now rank third in size of budget and account for more than 7 percent of a $60 billion dollar USDA budget (Figure 3.3). Food safety, although not a large budget item, grew in policy importance in the 1980s. Farm program expenditures, while varying widely, continued to decline slowly as subsidies have been constrained or reduced in every farm act since the 1985 legislation. This trend ended in 1998 when Congress began responding to another farm income "crisis."

The politics of agriculture became even more fragmented over the 1980s due in part to continued industrialization of the food system, and also to an opening up of access to U.S. political institutions, including Congress, and to a progressive fragmentation and proliferation of all economic and social interests (Browne, 1995; Bonnen et al., 1996). Agricultural interests and leaders in Congress can no longer exclude other voices from agricultural policy, and agriculture, as an economic sector, has become too interdependent with other domestic sectors and too important and politically difficult a dimension of trade negotiations to be left to its own devices in legislation and trade matters. Thus, depending on the topic, other cabinet agencies now participate in and influence decisions that once were the sole purview of USDA, or nearly so: the Agency for International Development and the Department of State in foreign food aid, the Department of the Interior and the Environmental Protection Agency in many environmental and natural resource issues in agriculture; and the Department of Health and Human Services in food programs. Half the cabinet participates in international trade negotiations, typically including the departments of State, Treasury, Commerce, Labor, and Defense. The policy process has become far more complex and deci-

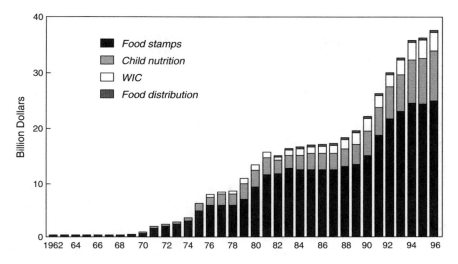

FIGURE 3.4 Federal expenditures on food programs, 1962-1996. Source: Economic Research Service, Historical Budget Outlays, electronic data base, U.S. Department of Agriculture, March 1995.

sions more difficult to reach. At the same time, the "farm vote" is now so relatively small that, unless the farm sector is in serious trouble (as it is again today), presidents no longer feel compelled to pay much attention to farm policy during their campaigns—or afterward. Congress is the dominant forum for farm policy today.

In ERS, John Lee followed Kenneth Farrell as administrator in 1981 and served for 12 years through the Reagan and Bush administrations. He retired in 1993, having had the longest tenure of any administrator of ERS. Lee and his first boss, Assistant Secretary for Economics William Lesher, inherited political trouble with Congress. ERS had done earlier research on the recent embargoes, which suggested that they had failed to achieve their purpose—a conclusion that a later, large-scale, academic-based research project validated (Economic Research Service, 1986). This result, plus the conclusions of work in the 1970s on the growing concentration of production and the changing structure of farming, left some influential members of Congress quite critical of ERS. One member of Congress introduced an amendment to the 1981 Senate agricultural bill that would have reduced ERS numbers by more than 100 positions. The new, politically experienced leadership of the department successfully defended ERS (Lesher, 1991). The assistant secretary subsequently had other uncomfortable occasions to defend ERS research from attack, including pressure to kill an ERS research project.

Over the 1970s and 1980s, many farmers became increasingly resentful of the growing intrusions of nonfarm interests and programs into "their department" and "their programs." This resentment created political difficulties for Congress and USDA as early as 1973, when the negotiations with labor and consumer interests needed to pass the 1973 farm bill had to be handled through third parties to protect the principals from political retribution (Bonnen, 1980). Thus, when the farm crisis of the early 1980s occurred, not only was a substantive response to farmer problems necessary, but also a less substantive—even symbolic—response was needed to assure farmers of USDA political concern and commitment to action.

In this context and under the pressure of both an erosion in real dollar resources and a decline from 1970 to 1980 of well over 100 personnel positions, ERS made two organization changes. The field staff, stationed at land grant universities since the BAE days, was disbanded in 1983 and its approximately 200 positions moved to Washington, D.C. (Rasmussen, 1991:89). This had a substantial negative impact on the smaller departments of agricultural economics and on ERS-university relations. From the Washington, D.C., point of view, the field staff had become too involved in university research agendas and was disconnected from the national mission of ERS. Field staff, in contrast, saw their role as one of helping to maintain good university-ERS relations, providing ERS with direct access to university research, facilitating joint work and also creating greater ERS capacity to understand and analyze local and regional problems, which the growing complexity of the farm programs required ERS to address. A significant number of the field staff left ERS rather than move to Washington, D.C. This increased the size of the staff in Washington, D.C., but it did not change total ERS numbers, since field staff had always been included in the total personnel numbers. Total ERS personnel numbers have continued to decline up to the present (Table 3.2). Following the substantial increase in their work on farm crisis problems, ERS in 1987 again modified its division structure to accommodate the focus of its work, changing division names so that it appeared symbolically as if agriculture were its only concern—which clearly was not the case (Box 3.1).

Changing Resource Base

With a few fluctuations, inflation-adjusted appropriations for ERS grew by 24 percent between fiscal 1962 and fiscal 1979, and the number of employees remained relatively stable, at about 1,100 until the mid-1970s. Since 1978-1980, inflation-adjusted appropriations and personnel numbers for ERS both trace a pattern of slow decline, to just over a 20 percent drop in the budget and more than a 40 percent cut in personnel. Almost 60 percent of this decline in personnel and almost 80 percent of the decline in budget have come since 1992.

In addition, ERS received budget transfers from other agencies. These transfers came both from within USDA and from outside, for contract research and analysis (Table 3.1). In inflation-adjusted terms, these transfers have ranged from a significant part of total resources (26 to 29 percent) in the last half of the 1960s to a quite small proportion (4 to 5 percent) over the late 1980s. In the period since 1992, the importance of transfers has increased again, varying between 10 and 18 percent of ERS's total inflation-adjusted resources.

One has to add the qualification that some unknown portion of such transfers in the USDA budget have periodically been pass-throughs to land grant universities and other organizations and not net additions to ERS resources. The Agency for International Development (AID) financed much of ERS's early work in international food aid and development. AID has continued to depend on ERS research capacities, most recently contracting $2-3 million per year with ERS from 1992 to 1996. The Soil Conservation Service (now the Natural Resources Conservation Service) has funded many major research efforts in ERS, including the large multiple river basin studies of the 1960s and the 1970s (Bowers, 1990:242-243). A good portion of the increase in budget transfers since 1992 has consisted of pass-throughs to state organizations from the Federal Crop Insurance Corporation ($1-2 million per year) and the Rural Housing and Community Development Service ($2-3 million per year). Thus, the apparent increase in budgetary transfers to ERS since 1992 is illusory, since most of the increase appears to be in the form of pass-throughs (ERS Information Services Division communication to panel).

When ERS and the Statistical Reporting Service (now NASS) were established in 1961, each had about 1,000 employees. In fiscal 1998, ERS had 554 and NASS has about 1,100 total personnel (Table 3.2; NASS internal communication). In each case in the 1960s, professionals probably constituted no more (and probably less) than half of the total number of employees. It is important to remember that, in a computationally intensive environment during the age of the preelectronic office, most calculations—both research and statistical—were done by large pools of clerks using technologies no more advanced than pencils, tabular paper, slide rules, and comptometers (precursor to the early Frieden and Marchant calculators). In these early days (Seaborg, 1991:25-26):

> You could see ditto machines with the smell of stencils and black ink, not copy machines that can readily produce many collated copies of your project. Slide rules were in common use as office machines made a lot of noise because they were mechanical. IBM electric typewriters were typical, but there were still many manual machines in use. Stacks of punch cards filled file drawers because that is how data was entered into the mainframe computers. By-the-way, the drum type computer was still in use. The use of computers was carefully monitored and each job order had to be justified. It often took several days to run a rather routine analysis.

Information on the composition of ERS personnel is available for only a few recent years (from 1995 on). Between 58 and 60 percent of all ERS positions

TABLE 3.3 Economic Research Service Permanent Staff

Count at end of the fiscal year	Economists	Other Social Scientists	Other Staff	Total Staff
1995	354	26	231	611
1996	349	22	213	584
1997	331	21	202	554
25 April 1998	306	19	198	523

Source: Information Services Division of the Economic Research Service, U.S. Department of Agriculture.

today are held by professionals classified as economists. The next largest group, 36 to 38 percent, are support staff. Another 5 percent of the ERS staff are social science professionals other than economists (Table 3.3). Of the economists, 62 percent have Ph.D.s, 35 percent have master's degrees, and 3 percent have bachelor degrees (Table 3.4). Although actual counts are not available, estimates obtained in interviews with ERS economists from the 1960s suggest that, given the office and computational technologies of the early 1960s, at least half, probably more, of early ERS personnel were support staff, whereas today the number is between 35 and 40 percent. If these estimates are accurate, then the numbers of professional economists in ERS have declined by 40 to 50 percent since 1962 and support staff by about 65 percent.

It is also worth noting that in the BAE and in the early ERS days, all, or practically all, economists on USDA's rolls worked in the BAE and ERS. Today, however, the majority of USDA economists are employed outside ERS in other USDA agencies, most significantly in the Foreign Agriculture Service, the Natural Resource Conservation Service, and the Forest Service. In addition, the larger action agencies now typically have a dozen or so of their own professional economists working on their economic policy and regulatory program issues (Table 3.5).

The decline in ERS resources and personnel can be attributed to a number of

TABLE 3.4 Highest Degrees Held by Economic Research Service Economists (March 1997 Survey Data)

Total Respondents	Ph.D.	Master's	Bachelor's
297	183	104	10
Percent of total	62%	35%	3%

Source: Information Services Division of the Economic Research Service, U.S. Department of Agriculture.

TABLE 3.5 U.S. Department of Agriculture
Employees Classified as Economists (as of 13 May
1998)

Economic Research Service	306
Foreign Agriculture Service	113
Natural Resource Conservation Service	66
Forest Service	47
Rural Business and Cooperative Service	26
Agricultural Marketing Service	25
Office of the Chief Economist	18
Farm Services Agency	16
Animal and Plant Health Inspection Service	12
Others (9 agencies)	29
Total	658

Note: There are many more economists in the USDA, a significant
portion with M.A. and Ph.Ds, who are classified in other categories.
A major example would be the large number of "foreign service offic-
ers" in the Foreign Agricultural Service, most of whom were trained
as economists.

Source: Information Services Division of the Economic Research
Service, U.S. Department of Agriculture.

sources. Over the period since the 1960s, automated and electronic equipment
have substituted for a good portion of the statistical clerks and other support
personnel. After a period of growth in government-wide investment in social
science research over the 1960s and 1970s, the decade of the 1980s saw major
cutbacks in such research and the growth of an anti-intellectualism in society.
Support for government research has eroded, especially in highly visible, contested areas of public policy.

In the more recent period since fiscal 1992, government efforts to reduce the
large federal budget deficit have resulted in declining appropriation for most cabinet agencies. In addition, interviews conducted by the panel produced multiple
accounts of dissatisfaction with the quality of ERS analysis and its responsiveness to clients. These are difficult to assess. Complaints ranged from erroneous
analyses, to refusal to share research done for trade negotiations in agriculture, to
the belief that some critical research and analytic product provided to Congress or
the Office of Management and Budget was constrained by the politics and policy
positions of USDA. These complaints were often contested and may be wrong,
but it is important to understand that irrespective of the complete accuracy of
these complaints, these perceptions alone erode the credibility of ERS and the
value of its products.

It should be noted that since 1992 the ERS base budget has declined at twice
the rate of USDA as a whole. If executive and legislative leadership do not soon

agree on common rules for access to ERS services and act to ensure the performance, independence, and credibility of this research and information function, it will continue to decline in capacity and relevance to the increasing complexity of the industry and policy area it serves in food, agriculture, natural resources, and the environment.

LESSONS LEARNED

More than seven decades of BAE-ERS experience suggest some clear principles that govern the successful organization and management of a federal agency dedicated to research and analysis of high integrity.

First, the history of the BAE-ERS demonstrates the important value of a cadre of professionals in government capable of providing reliable primary data, secondary data and analysis, short-term problem analysis, and longer-term research products—especially when faced with new problems or in conflicted periods of rapidly changing conditions (e.g., see Finegold and Skocpol).

Second, a federal research and analysis agency must be protected by both its cabinet secretary and the Congress from politicization of its products and its internal decisions. The integrity and long-term survival of such a service function is highly vulnerable to the political conflict that accompanies the policy process. Government research units have little capacity to protect themselves through their own actions, and few of their clientele are willing or able to do so. In turn, the agency must expect, and be allowed, to provide both the Congress and the secretary with access to reliable, high-quality research and information products (Kirkendall, 1982:257; Bowers, 1990:236). Failure of the agency to provide a high-quality, relevant product will lead to the eventual demise of the agency. As a corollary, Congress and the cabinet secretary must be free, and be expected, to disavow any responsibility for a research or information product that conflicts with their views, even as they must defend the independence of the research agency.

Third, all major federal research and information products eventually develop a diverse set of users in both the private sector and well beyond the federal government in the public sector. In a democratic, open society, this is both inevitable and necessary. Common rules should govern such access, if conflicting expectations are to be avoided.

Fourth, if it is to maintain a reputation for, and the reality of, integrity of product, a federal research and analysis agency must not be expected to participate directly in policy decisions. It may evaluate problems and alternative solutions, but it may not safely recommend or advise a specific policy action or participate in its implementation. As we have seen, such actions can destroy an agency. As a consequence, as both T.W. Schultz (1954) and John D. Black (1947:1035-1036) long ago concluded, the day-to-day staff work for the secretary's office and the other "constant, routine, 'trouble shooting' work or quick program analysis" (Schultz, 1954:20) should not be done by a federal agency responsible for objective, high-integrity, longer-term analysis and research.

Fifth, a federal research agency should be organized and managed to take advantage of the strong complimentarity between research products, problem analysis and secondary data development, and its analytic structures (Schultz, 1954:20-21).

Sixth, the Office of the Secretary must have the capacity to use economic research and problem analysis. Since it is unrealistic to expect every secretary to be an economist, this means that there should be a political appointee (i.e., assistant secretary for economics or chief economist) who has these skills and who reports directly to the secretary.

A federal research unit should be independent and protected by the secretary's office from all direct political influence, but this arrangement creates an organizational dilemma. It must also be connected to the line structure of the cabinet department at high enough level to remain well informed about the longer-term information and research needs of decision makers but remain functionally separate—i.e., independent. Lessons from the past strongly suggest that the research agency should report to an assistant secretary for economics (or policy). This would be a political appointee with direct access to the secretary who is a policy economist with experience necessary to understand, protect, and balance the goals of independence, policy access, and responsiveness (Cochrane, 1983:30). History demonstrates that failure to provide such a position leaves the research unit subject to the political whims and pressures of cabinet secretaries and deputy secretaries who are often under too much political pressure to respect the integrity of such units. Since 1993, ERS has been in just such a vulnerable and hazardous position within the line authority of USDA.

4

Current Issues and Problems

The mid-1990s have been a period of continued economic growth following the brief recession in 1991-1993 in which growth slowed and unemployment grew. The real value of the dollar, which had been declining since 1985, stabilized and from 1992 to 1996 fluctuated around the same level (Council of Economic Advisers, 1998:247-248, 330, 408). Since then, major economic problems and consequent devaluations of currencies in Asia, Russia, and elsewhere have led to a stronger dollar, with the expected depressing effect on U.S. exports, especially in agriculture.

World trade has been expanding faster than world economic output since 1973 (Council of Economic Advisers, 1997:243-244). This is a direct consequence of the growth of open national economies and the globalization of financial and commodity markets. The prosperity and growth of most nations now depends on the expansion of trade to an unprecedented degree. The U.S. agricultural sector is even more dependent on trade. In the mid-1990s, exports accounted for about 30 percent of U.S. agricultural output. Typically the United States exports the production from about half of its wheat acreage, one-third of its rice, soybean, and cotton acreage, and one-fifth of its acreage of feed grains. Agricultural exports reached $60 billion in 1996, but the financial difficulties of Asian, African, and Eastern European countries reduced demand for U.S. exports in late 1997 and 1998 (Council of Economic Advisers, 1998:397; Economic Research Service, 1998a). The current financial problems in developing countries are likely to slow the growth of U.S. agricultural exports at least through the end of the decade. The long-term prospects for increased food exports depend on world economic growth and on the continued expansion of trade, especially as it involves in the low-income developing nations of the world.

In the late 1990s, the United States appears to be entering another major farm crisis. Economic disorder in Asia, Eastern Europe, and Russia has rolled through international capital and commodity markets. The exchange value of a number of currencies fell and was often followed by devaluation and government control of foreign exchange. This and the resulting internal inflation has cut effective demand for U.S. exports and driven the value of the dollar in trade with these countries to far higher levels, further cutting U.S. export demand.

In the United States, a large carryover, especially of food and feed grains, combined with near-record grain and soybean production in 1998, had already depressed farm prices. The subsequent rapid decline in U.S. farm exports pushed farm prices to near depression levels. The U.S. response in a congressional election year has been large increases in emergency farm income support. Since the carryover of stocks into 1999 will be even greater than into 1998 and recovery in export demand cannot be expected soon, the price and income problems and their consequences are likely to continue into the near future.

The major difference between the current crisis and that of the 1980s is that farmers are carrying lower levels of debt today and fewer are as highly leveraged as they were in the 1980s. Despite the massive changes in the food and agriculture sector, one sees again a continuing characteristic of the classic farm problem—a persistent disequilibrium in the form of an excess production in the face of ruinous prices. Farmers tend to maintain excess productive capacity in the same products primarily because of large investments in specialized assets that have no value in the production of other crops or livestock (Kilman, 1998).

The previous chapter, on the origins and history of the Economic Research Service (ERS) and its predecessor agency, sets the stage for our discussion here of the ERS in the 1990s and beyond. We begin by characterizing the enormous changes in the world that have affected and are continuing to affect the agricultural sector and the agricultural policy agenda in the United States. We then bring the history of ERS up to the present by describing the difficult years of the past half decade, providing as well our vision of the likely concerns in the foreseeable future. The remainder of the chapter lays out the most important issues and problems facing the U.S. Department of Agriculture (USDA) and ERS today.

CONTINUED GROWTH IN THE COMPLEXITY OF POLICY

The pattern of agricultural legislation over the entire period since World War II has been one of increasing complexity. The simply titled Agricultural Act of 1948 was only 13 pages long. In 1973, 25 years later, the act was still only 29 pages. By 1990, the Food, Agriculture, Conservation and Trade Act required 713 pages (Bonnen et al., 1996:147). It was a document of such complexity that many commercial farmers needed years of experience with program regulations plus the help of an accountant and a lawyer to make well-informed program participation decisions. The 1996 Federal Agricultural Improvement and Reform

Act (FAIR) initiated a major change in farm policy: it eliminated all production controls, set a fixed annual subsidy that declines each year, and is operative through the year 2002. Whether this policy approach can be sustained through 2002 or beyond is now at issue (Kilman, 1998).

Although the general direction over two decades in farm policy has been toward letting farm markets work with less direct government intervention, the growing number of policy participants and multiple policy goals (high farm income, stable prices, agricultural competitiveness, access to world markets, environmental sustainability, reduction of budget deficits, etc.) have led to more and more conflicted and complex legislation. Policy analysis requirements, as a consequence, have also grown more complex and demanding.

There has been a steady march since the late 1960s toward interdependent, open national economies tied together by global financial and commodity markets. Global markets now dominate trade in most agricultural and natural resource products. Closed economy models of national economic policy problems are almost sure to mislead. This is especially true of sectoral policy, such as in agriculture. Research strategies and agendas must now be cast within a macroeconomic framework.

From the 1930s into the 1970s, the possibilities for agricultural trade policy were bound by the constraints created by domestic farm policy. Since the 1980s, the reverse has been true. Domestic farm policy has become constrained by increasingly open and competitive world markets and in the 1990s by the Uruguay round of trade agreement commitments under the General Agreement on Tariffs and Trade (GATT). Today the major challenge to the income protection interests of production agriculture arise from federal budget constraints and from U.S. treaty commitments to reduce agricultural subsidies and protection. The political and policy context for USDA and thus for ERS has changed significantly. In addition, during the 1990s, food safety has become an important issue with an apparent increase in food-borne health threats. Environmental and natural resource policy issues now involve growing public and scientific concern, to which increasing political attention is paid. These highly political issues raise regulatory policy questions, in which economic research and policy analysis have an important role to play. It is especially important to explore the efficiency of economic incentives versus direct regulations. The alternative to ex ante research and analysis is large, poorly informed, expensive, and largely uncontrolled policy experiments on the economy and the body politic.

The USDA has become programmatically more complex. Its budget now runs over $60 billion a year, about 60 percent of which goes to food stamp recipients. Only about 20 percent of the USDA budget is spent on farmer-related programs and activities. And 40 percent of the more than 100,000 USDA employees work for the Forest Service. The diverse programmatic structure of the department is reflected in the fact that the expenditures of USDA now fall into 10 of the 17 budget categories of the Office of Management and Budget (OMB), more than any other cabinet department (Madigan, 1992; Office of Management and Budget, 1998a).

It is not surprising that ERS publications in the 1990s reflect an ever-increasing diversity of topics. ERS is now called on more frequently to serve the department's needs in its policy interactions and negotiations with other cabinet agencies and units of state government. Publications reflect this in analysis of such topics as water policy in the Pacific Northwest, coastal zone management, reduction of pesticides in foods, changing food stamp rules, and the economic adaptation of agricultural production to global warming. The number of publications concerned with global market issues has grown. By the mid-1990s, the clientele of the USDA included 1.8 million farmers, about 25 million food stamp recipients, 25 million children and 5 million single adults in supplemental food and nutrition programs, agribusiness firms, timber companies, environmentalists, and rural communities.

The pressures on ERS to expand its scope of work run in two politically inconsistent directions. In the 1990s, problems of production agriculture and its natural resource base have been fragmenting into many new and complex issues involving traditional and politically influential clientele who are demanding more attention. At the same time, many other, increasingly important new areas of policy concern are proliferating, along with political pressures for action from politically influential interests. This occurs, for example, in the case of global warming, food safety, health and nutrition, low-income family and child nutrition and access to food, new regulatory problems, rural community decline and development, environmental issues in ecosystem sustainability, water and air quality, toxic chemical use, hazardous wastes and endangered species, and on and on. With limited and declining resources and little capacity to resolve these conflicts over priorities, ERS is being blamed for not satisfying all claimants on its services.

Difficult Years for ERS

A lack of stability has seriously affected ERS in the 1990s. Between 1990 and 1998 USDA has had four secretaries, only one of whom has served much more than two years: this includes Clayton Yeutter (1989-1991), Edward Madigan (1991-1993), Mike Espy (1993-1994), and now Daniel Glickman (since 1995). Since 1993, when ERS administrator John Lee retired, ERS has had four administrators, the first three of whom were "acting" and thus temporary (Box 4.1). For much of this period, not only was the secretary relatively new in his tenure, but also congressional pressure for a major reorganization of the USDA was occupying major time and political energy of the secretary's office. This politically difficult reorganization involved combining action agencies with grassroots political influence into one national headquarters and closing or bringing their local offices into common locations (Office of the Secretary). Attention to ERS's need

BOX 4.1
Leadership in the Bureau of Agricultural Economics, the Economic Research Service, and the U.S. Department of Agriculture

Chiefs of the BAE 1922-1953

Henry C. Taylor	1922-1925
Thomas P. Cooper	1925-1926
Lloyd S. Tenny	1926-1928
Nils A. Olsen	1928-1935
Albert G. Black	1935-1938
Howard R. Tolley	1938-1946
Oris V. Wells	1946-1953

U.S. Secretaries of Agriculture 1945-1998

Clinton P. Anderson	1945-1948
Charles F. Brannan	1948-1953
Ezra Taft Benson	1953-1961
Orville L. Freeman	1961-1969
Clifford M. Hardin	1969-1971
Earl L. Butz	1971-1976
John A. Knebel	1976-1977
Bob Bergland	1977-1981
John R. Block	1981-1986
Richard L. Ling	1986-1989
Clayton Yeutter	1989-1991
Edward R. Madigan	1991-1993
Mike Espy	1993-1994
Daniel R. Glickman	1995-

for permanent leadership was not paid until 1996, when the current administrator was finally appointed.

As a part of continuing spending reduction efforts, the Office of Management and Budget in 1993 proposed to cut ERS funding for fiscal 1994 by a draconian 25 percent. At the time the administrator of ERS expressed the belief that "a cut of that magnitude will be supported by the department and by Congress" (Lee, 1993a). Such a cut would have required a huge personnel "reduction in

Administrators of ERS 1961 to Date

Nathan M. Koffsky	1961-1965
Melvin L. Upchurch	1965-1972
Quentin M. West	1972-1977
Kenneth R. Farrell	1977-1981, ESCS Administrator
J.B. Penn	1977-1981, Associate Administrator for Economics
John E. Lee	1982-1993
Katherine Reichelderfer (acting)	1993
Kenneth L. Deavers (acting)	1993-1995
John C. Dunmore (acting)	1995-1996
Susan Offutt	1996-

Directors of Agricultural Economics and Assistant Secretaries of Agricultural Economics

Willard W. Cochrane	1961-1964
John A. Schnittker	1964-1965
Nathan M. Koffsky	1965-1966
Walter W. Wilcox	1967-1968
Don Paarlberg	1969-1977
Howard N. Hjort	1977-1981
William G. Lesher	1981-1985
Robert L. Thompson	1985-1987
Ewen M. Wilson	1987-1989
Bruce L. Gardner	1989-1992
Daniel A. Sumner	1992-1993 (position eliminated in 1994)

force." This appears to reflect a then-prevailing dissatisfaction with ERS, including a perceived lack of responsiveness to clientele, and an alleged uneven quality of analysis, including failures to understand the full context of some policy issues being analyzed.

More than two-thirds of the reduction in ERS budget from its high point in 1979 has come since 1992 (see Table 3.1). Total ERS personnel numbers have declined by more than half from their historic peak in 1968, 37 percent of that

decline having come in the brief period since 1992, when the mandated downsizing began (see Table 3.2). To meet reduced personnel ceilings three early
retirement buyouts occurred, further adding to the disorder in ERS during this
period. Symbolic of the downsizing of ERS's professional research and analytic
capacity during the 1990s was the 1994 demise of *Agricultural Economics Research*, a peer-reviewed professional journal established by the BAE in 1949 and
published by its successors for 46 years. In addition, during the 1994 reorganization of USDA, the position of assistant secretary for economics (to whom ERS
had reported) was eliminated. ERS was moved to report to the under secretary
for research, education, and economics (Figures 3.2 and 4.1). ERS accounts for
only 3 percent of the 1997 budget outlays for which the under secretary is responsible (Office of Management and Budget, 1998b, 1998c). Since biological science research accounts for most of the budget and activity for this under secretary's jurisdiction, over the long run the position is most likely to be filled by a
biological science researcher. In today's specialized world, the odds are against
such a person's having broad policy experience or much understanding of economics.

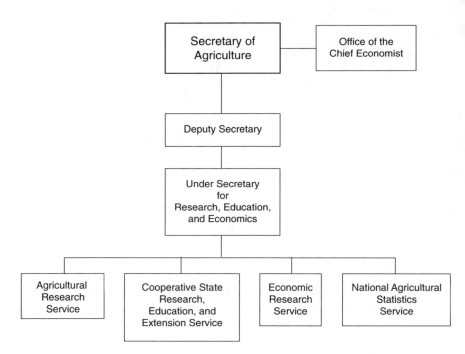

FIGURE 4.1 Reporting line for the Economic Research Service (ERS), 1994-1998.
Source: Office of the Secretary, Memorandum 1010-1 of October 20, 1994.

In the midst of this serious uncertainty and downsizing, acting Administrator Kenneth Deavers (1993-1995) managed a divisional reorganization. Subsequently, current Administrator Susan Offutt, also faced with downsizing, reallocated personnel and budgets in 1997 to reduce the number of divisions from four to three. The large shifts in priorities and the reward system over the 1990s have left ERS personnel confused. Many now wonder whether any of the broad categories of ERS products, objective economic research, policy and other analysis, or even basic information provided by secondary data (including *S&O* work) are still valued by the USDA—to say nothing about concern for an appropriate balance between such complementary outputs. The net effect on ERS of events in the 1990s has been confusion, low morale, and a lack of clarity as to what the department or Congress expects of ERS.

Future Direction of Policy

The demand for ERS research, analysis, and secondary data and information services depends on the policy problems and agenda of the department and the economic and societal sectors it serves. To the extent that one can anticipate the direction that future policy is likely to take, one can envision the various kinds of products ERS should be prepared to provide.

The nature of the agriculture sector is evolving rapidly. The organization of both production and farm input and farm product markets continues to proliferate into many separate production and market structures with different economic characteristics. This proliferation creates an increasingly diverse policy setting. Some parts of agriculture are being integrated into the chemical industry as sources of biological feedstock. Transgenic animals are beginning to be used to produce pharmaceutical products. Products and their markets are being differentiated and organized by consumer preferences. Biotechnology makes it possible to redesign a crop to fit specialized consumer preferences and different end uses. Some farmers no longer raise corn or soybeans, but rather a specific genetic type of plant for a specific end use.

The growing number of genetically engineered crops, animals, and animal products are raising new issues. Questions about the safety of these new products for consumers and for the environment must be evaluated or otherwise dealt with. The current rapid concentration and contractual coordination of agricultural production and marketing firms, both vertical and horizontal, is now accelerating, especially those specializing in biotechnology-based products. The U.S. Patent Office received 4,000 patent requests for nucleic acid sequences in 1991 and 500,000 in 1992 (Enriquez, 1998). Seed, chemical, pharmaceutical, and biotechnology firms are merging to form highly concentrated new industrial structures (Service, 1998; Enriquez, 1998). Is there a significant public interest in industries in which monoploid concentration occurs? Clearly many of these biotechnology innovations will substantially increase the yields and output of some farm products. On one hand, will the innovations in any of these product markets compound the classic farm problem of income depressed by steady increases in output in the face of inelastic demand for the product? On the other hand, have improved information, new technologies, and new institutions in some markets offset the forces of the classic farm problem in some markets, eliminating the need for government intervention? Certainly there is no suggestion in current agricultural price and income problems that asset fixity or specificity of asset use in farming and its destabilizing consequences have disappeared (Kilman, 1998).

Agriculture and agricultural markets, once organized around fairly standard bulk commodities, are breaking up into many different products, production processes, and differentiated markets. Increasing numbers of agricultural markets face global competition. Information technologies and changes in public regulations are facilitating a massive reorganization of agriculture. Previously settled property rights are being modified, especially by the biotechnology and information technology revolutions and by adjustments to globalization. Although it is hard to tell where these changes will end, the road ahead is bound to be a bumpy ride for farmers, farm- and food-sector firms, and those with public-sector responsibilities.

It especially has to be appreciated that the rate of change in the food system is accelerating rapidly. Also, this transformation is not limited to the farming sector but includes the entire food system, from farm input industries, farm product processing and manufacturing, to food marketing through the retail level. What the structure of the world food system will eventually look like and who will control it, or its various components, are not clear at this point. What is clear in the United States is that the traditional agriculture of many independent farms dealing with competitive farm product processing and marketing firms, all coordinated by a relatively transparent set of open domestic and international markets, will be a far smaller and shrinking part of the food system.

Coordination of the newer evolving structures is likely to be dominated by direct vertical and horizontal integration and control or by contracting arrangements or some mixture of direct integration of functions and contracting. Many of the integrated firms or contractual systems will be global in scope. Indeed, many food system firms are multinational now. What were once open markets will in many cases disappear or become perilously thin. A large part of food and agricultural market information will become proprietary or unreliable and difficult to retrieve for other private or public uses. Despite these difficulties, comprehensive production, price, trade, and finance statistics will still be a necessity for public policy and private business planning. The prospect for the short term of two to three years is one of great uncertainty. The longer run of five years and beyond is seen as one of far greater uncertainty, as industry analysts and participants anticipate a tidal wave of changes in technology, industry structure, and financial control, as today's already changing food system is pulled apart and reorganized in entirely new ways. In turn, public policy for the food and agriculture system has not encountered the uncertainty it now faces since the Great Depression, when most of the institutional and policy framework of the last two-thirds of a century was put in place.

The trend in U.S. agricultural and trade policy toward reduction in market protection and direct subsidies is likely to continue as long as trade expansion remains a dominant policy objective. This trend does not mean that eventually there will be no public policy applied to many of these differentiating markets—as some policy participants and analysts seem to expect. Instead of price and production controls and direct subsidies, a heterogeneous mix of various regulatory regimes is likely to prevail. Some subsectors of agriculture will become much like other parts of the industrial and commercial economy that experience little or no direct government control. However, markets do not exist without rules, both public and private. Even if the traditional farm programs come to an end in 2002, market participants, depending on their problems, are almost sure, like those in other economic sectors, to lobby for or against other types of regulation—e.g., setting standards, modifying the tax code, and changing property rights and market rules. Just as in other sectors, there will be regulatory efforts to extract subsidies and create protections that are not as visible and vulnerable as direct budget subsidies and price supports. This greater complexity of the food and agriculture sector will create more, not less, demand for policy research, analysis, and secondary data. The emphasis will be on analysis of business practices and complex, less transparent regulations rather than the direct price and production interventions of the traditional farm programs.

Agricultural, food safety, natural resource, and environmental problems and regulation are no longer domestic but rather international policy issues. Capital and financial markets and many commodity markets are now global in scope. Because of the ease and speed of transportation and communication today, border controls can no longer enforce domestic policies as effectively as they once could. Political pressure grows for internationally agreed-on common policies or regulations (Cohen, 1998). Some former national powers and policy responsibilities are moving to the international arena, as others are moving to subnational levels of government and to private-sector institutions. Old institutions are being modified and new institutions created (Guéhenno, 1995; Mathews, 1997; Rodrik, 1997). International trade treaties have become an omnibus vehicle for achieving multiple policy goals in agriculture, labor, natural resources and the environment, health and food safety, and other areas. This internationalization is leading to the creation of international regulatory institutions and to a greater role for international organizations in policies and policy making (Schuh, 1991; Bonnen et al., 1997). All of these changes contain dimensions requiring research and policy analysis for understanding the problems as well as creating needs for many new categories of secondary data and information.

The future for USDA and ERS appears to be even more difficult to deal with than the present. Until new national and international institutions of market and public governance evolve to order this complexity, greater political and economic uncertainty will prevail. The optimal design of institutions is an important and researchable question. Both on the domestic and the international scene, we appear headed toward regulatory policies that in subject matter and scope of regulation will be more complicated. The institutional structure to implement policy is likely also to be more of a maze, with problems of overlapping areas of authority among national as well as international regulatory bodies. USDA needs to ensure its capacity to deal with this new world. Much as when it faced open-ended uncertainties in the Great Depression, USDA will, if it is wise, turn to its economic and social science research and analysis capacity for support in understanding and contending with its responsibilities.

Agricultural commodity production is by nature location-specific, involving millions of different ecosystems and greatly differing conditions of production. Nevertheless, worldwide competitive pressures on many older production regions are growing due to (1) newly developed areas of production, (2) the recovery of agricultural productivity in some former command economies, and (3) the impact of biotechnology and other technological and institutional innovations that lower unit costs or create substitutes. The processing of commodities does not have this location specificity and thus can be located anywhere that transportation facilities, resources for processing, and access to markets are available. Nations, including the United States, cannot assume in their policy making that there are no alternative locations in the world for the further development of their natural resources beyond the primary production level. When markets are increasingly global, so is competition.

The need for research and policy analysis on natural resource and environmental problems will continue to grow, as worldwide population and economic growth put additional pressure on the natural resource base and the environment. Responding to this pressure will be in some part a USDA responsibility, since the natural resource base of the United States, as elsewhere, is located mostly in rural areas and is, for public policy purposes in the United States, the responsibility of the departments of Interior and Agriculture and of the Congress.

Research on consumer interests in food safety, nutrition, and related food matters are also likely to become more important. This increasing importance is inevitable as consumer preferences dominate and differentiate domestic and international markets as never before. As many of these markets become global and highly coordinated or vertically integrated and concentrated, market performance, anticompetitive practices, and antitrust issues will be raised.

ISSUES AND PROBLEMS FACING ERS AND USDA

Today ERS faces a series of problems, some persistent dilemmas, and other more recent challenges. ERS can resolve some, but there are other problems that only the secretary of agriculture and the Congress together can address.

Balance Among Products

ERS, and the Bureau of Agricultural Economics (BAE) before it, have produced three overlapping, highly complementary, and often integrated products: (1) intermediate and long-term research, (2) policy analysis, and (3) with the National Agricultural Statistics Service (NASS), the basic information and intelligence base for policy decisions in food, agriculture, and rural resources, both public and private. ERS's contribution to this third broad category is not basic data collection and its initial processing, which is the role of NASS, but rather secondary data development and related analysis. This ERS product, sometimes called value-added information, involves combining data sets from diverse sources and processing such data through various kinds of analytic models and accounting systems to produce indexes, income and other financial estimates, and different types of projections and forecasts. Perhaps the most visible of these secondary data and analysis products are various ERS indicator series and the many *Situation and Outlook* (*S&O*) reports on different dimensions of the U.S. and global food system.

Since the early days of the BAE, it has been clear that there is little or no consistent demand or any associated political support for long-term research. Such research is the necessary—but generally unrecognized—foundation for high-quality, shorter-term applied analysis of problems and policy questions, and for the development of new or revised secondary data and its analytic structures. The other foundation for good-quality policy analysis is current market intelligence and related institutional knowledge, including policy and political behaviors. It is clear that some secretaries of agriculture and some administrations have valued ERS policy and problem analysis, but others have not. The only consistent demand for ERS products has been the broad public and private usage of basic information and intelligence on food, agriculture, and natural resources, including its packaging as various economic indicators and *S&O* reports.

This leaves ERS with the difficult problem of finding support for and, with increasingly limited resources, sustaining an appropriate balance among (1) intermediate and long-term research, which has not been supported well in either USDA or the Congress, (2) various types of problem and policy analysis, which is inconsistently supported by the Office of the Secretary and Congress, and (3) the secondary data and analysis function. The recent ERS attempt to reduce resources in *S&O* activities led to organized political opposition from private-sector clientele, USDA action agencies, and the Congress. It would appear that indicators, *S&O* intelligence, and related basic information are the only highly visible, broadly valued, and consistently supported ERS products. All three of these products are necessary to the quality of the others and to ERS's capacity to perform well. But there appears to be little recognition of this complementarity by clientele or by USDA. Both the BAE and ERS have faced this dilemma without the capacity or support to resolve it in any satisfactory fashion. Over the nearly four decades of its existence, some ERS administrators have understood better than others the complementarity of the three categories of ERS products and the necessity for support from the Office of the Secretary and the Congress.

Different Internal Cultures

Both experience and bureaucratic theory indicate that, to achieve clearly defined, singular goals, an organization can be most effectively managed in a centrally controlled hierarchical structure with a clear, formal chain of command. However, when goals are not clear or are subject to large uncertainties, a less hierarchical, more decentralized and informal decision process works far better. In ERS, the conduct of basic research usually falls into the latter category, whereas information products and services fall into the first, more typically bureaucratic, category. Thus, there has long been a dilemma faced in managing ERS, whether recognized by its leadership or not: it is the conflict between the hierarchical organization of government and the need for decentralized, more informal, interactive research leadership. The branch chiefs and division directors want to know when their lines are being traversed, but researchers just walk down the hall and exchange information across bureaucratic lines. The formal organization conflicts with the informal organization. Some administrators have encouraged and abetted this duality because it was productive, and some have pressed for more central control.

Shifts from decentralized research decisions to relatively centralized control over research have changed not only with leadership preferences but also with technology. Examples are the shift from decentralized typewriters and word processors to centralized word processors and from decentralized, unnetworked personal computers (highly decentralized) to desktop computers attached to one central computer (highly centralized) to networks among personal computers (capable of either).

The dilemma lies in the fact that the management system that is ideal to achieve good-quality research is not ideal for the information and systematic intelligence output of the agency—and vice versa. Good management of the policy analysis function would seem to lie in a mix between the two extremes.

Organizational Support

In the budget reduction environment of recent years, another long-time problem for ERS has become more intense. ERS must work its way in a highly political environment dominated since the 1930s by action agencies that are politically well-connected to organized clientele and the Congress. As Chapter 3 shows, government research and analysis agencies do not compete well for resources with strong clientele-based program agencies. In this environment, the question of who does or must support government research and analysis capacity, if that capacity is to survive, is relevant and, in the case of ERS, it appears to be urgent. Although the primary and most direct customers of ERS products are the Office of the Secretary and the Congress, historically it is clear that *consistent protection and institutional support of ERS have not been forthcoming from either.* One active policy participant and analyst has observed about support for ERS (Doering, 1991:19):

> The fact that ERS has no clear client base and client service relationship (other than the Secretary of Agriculture) robs it of a definite base of support and the assurance of a stable and safe agenda. Its clientele are going in different directions wanting different things.

The White House is a distant and unconscious though dependent user of ERS intelligence and policy analysis. The White House is too occupied by larger questions to be concerned about ERS. The Office of Management and Budget (OMB) in the Executive Office of the President is another matter. OMB has a conscious, direct stake in the quality of ERS products. Although it is dependent on ERS products, in recent years the executive budget for ERS has been cut far more deeply than for the entire USDA. This appears to reflect a dissatisfaction with ERS performance.

The action agencies of the USDA are in many cases suppliers of information to ERS as well as important users of all ERS products. But they either presume ERS will always be there for them or, as likely, are nervous over whether ERS research and analytic results will support or undermine their programs. Many, though not all, are quick to criticize results they do not like and tend to support ERS only in a negative way when ERS proposes, as it did recently, to stop providing some product or service the action agencies use regularly. That way, they suddenly discover that ERS is doing something of value. Many private-sector market and policy interest groups depend on ERS information and research products, but they also tend to behave in the same way. Similar behavior can be found among academics who are dependent on ERS products.

The agricultural economics profession as a whole is dependent on ERS research and information products and on ERS membership and participation in the American Agricultural Economics Association. Since the 1980s and the withdrawal of ERS field staff from university departments of agricultural economics, the high levels of interdependence and mutual support that characterized the period of the 1930s through the 1970s or so have all but vanished. With this change, conscious attention to and support of ERS among agricultural economists also seems to have declined.

The depth of cuts in ERS budgets in the 1990s need explanation beyond the effort to reduce budget deficits. Diverse reasons have been alleged for the apparent lack of support for ERS. The possibilities raised range from significant ERS performance failures, unrealistic and conflicting expectations of ERS, belief that ERS analysis was on occasions politicized by USDA, to a few instances of supposed denial by the secretary's office of access to ERS products. No matter the degree of validity or generality of these criticisms, they create a perception that must be dealt with. Otherwise ERS will have little chance of achieving the full potential of a federal research and analysis agency.

Three things are clear. First, ERS needs to work to improve its customer satisfaction and support—whatever the sources of its problems are. Second, the Office of the Secretary and the appropriate units of Congress need to establish a common set of rules and expectations to govern their and other user access to ERS products. Third, the appropriate future role of ERS, its product mix, and resource constraints need to be jointly assessed by USDA, ERS, and major users of ERS products.

Expanding Scope of Responsibility and Shrinking Resource Base

Over the life of ERS, an expanding USDA agenda of policy issues and problems has created dilemmas affecting almost every dimension of ERS's performance. Not only has the diversity of information and analysis demanded of ERS substantially increased, but also, over the entire period, its real resource and personnel base has seriously eroded (see Tables 3.1 and 3.2). The ERS potential solutions must either stabilize the resource base or define its mission more clearly and perhaps narrowly, if it is to regain the quality of performance of which it has been, and should be, capable.

ERS has faced many negative external influences and constraints on its performance over the years. During the 1990s, as several respondents observed, it has been "jerked around continuously" with little concern in USDA for the stability or integrity of its functions. In this disorder and conflicting expectations, many of its best young economists and even more of its most experienced professionals have left for more promising positions elsewhere in government, academia, or business. At the same time, recruiting either has been constrained by downsizing or has simply not succeeded in attracting the best candidates. Morale is low and the current role of the agency continues to lack clarity. Many of the professional staff do not believe their work on the different but complementary ERS products (research, various types of analysis, and secondary data and information) is or will be rewarded. Such a situation cannot be allowed to continue, if ERS is to perform well or even survive.

Over the years, it has often been alleged that the quality of ERS performance and that of its staff has declined. Such statements have been made since the 1970s but were then being measured against the record of the BAE of 20 to 30 years earlier. BAE economists were an intellectually dominant element in a very small agricultural economics profession. By the 1970s, both the quality and size of academic faculties and the number of Ph.D. programs in agricultural economics had tripled or more. The ERS of the 1960s and 1970s still had many of the BAE economists plus outstanding newcomers, but ERS, although the same size, had a smaller relative presence in a much larger, growing profession. It no longer dominated the profession, but it was not accurate to say that either the quality of the ERS staff or its performance had eroded. The ERS of the 1990s is an even smaller, though still significant part of the agricultural economics profession.

Today, there is widespread perception that the quantity and quality of ERS products are not what they should be, given the number of professionals in ERS and the size of its budget. These reservations about the capabilities of ERS are becoming more acute as ERS is asked to address an ever-widening range of serious economic issues.

Historically, ERS operated with a mandate that was heavily circumscribed, relative to what it is asked to do now. The professionals who could best meet this mandate were produced almost exclusively by graduate programs in agricultural economics. In turn, a career in ERS was an attractive opportunity for many of the best students in these programs. Relative salaries for academics were lower, and until the 1950s ERS dominated the market for recently trained PhDs in agricultural economics. Private-sector opportunities for agricultural economists were less plentiful. When agriculture was a sector that could not be ignored in national politics, a successful career professional in ERS could reasonably expect to have a significant impact on farm policy. Today a professional with these ambitions is likely to look elsewhere, as the responsibilities of ERS have grown and compete for impact on policy.

Today, ERS is asked to address a much broader scope of economic issues. For many of these issues, professionals who can best do the required work are just as likely (if not more so) to be the products of graduate programs in economics and some other social science disciplines, as of agricultural economics programs. Although ERS has broadened its recruiting efforts beyond the land grant universities over the past decade, its mix of staff expertise is still short of that needed to respond to the growing scope of issues facing USDA. Even more telling, salaries and other dimensions of career opportunities in the academic and private sectors are much more attractive than they once were. These two circumstances critically limit the ability of ERS to provide research, analysis, and information in emerging areas of responsibility that is both of high quality and is perceived to be of a quality, comparable to the best that could be obtained.

These difficulties in meeting the broadened mandate of ERS with the work of career professionals have been compounded by several other factors. First, it has been difficult for ERS, or any other agency, to reward their best professionals with rapid advancement, and at higher levels to offer substantial advancement that does not entail administrative responsibilities. Second, the fractious political environment in Congress, and between Congress and the executive branch, has meant that policy makers have intervened more heavily in ERS agenda and publication decisions. Third, both current and former ERS professionals perceive that what is expected of them has changed drastically, as ERS administrators and division directors, plus USDA political leadership, have come and gone. The best graduates of economics and agricultural economics graduate programs have choices that largely avoid these problems, in positions with the same or better salaries and opportunities for advancement. The best professionals in ERS have also had these opportunities, and many have taken them.

There is thus ample basis for concerns about the quality of ERS performance, in the future as well as today. But these concerns are also a reflection of the larger scope of new USDA policy and information needs embedded in an inconsistent and conflicting set of priorities held by a diverse set of new and old clientele, all colliding in the context of a progressively thinner ERS resource base. Academics, for example, cannot happily continue to criticize ERS performance, while discouraging their better students from considering ERS employment and still expect ERS to provide a high-quality, comprehensive, information, and analytical base for the profession—as it now does. The critical problem is that ERS, like USDA, is now trapped between the politics and political pressures of older, mostly agricultural clientele and newer USDA clientele and issue advocates. But ERS has little or no capacity to resolve the growing conflict over priorities for the use of its limited resources. That responsibility lies in the Office of the Secretary and the Congress. The future of the ERS information and analysis base is clearly at hazard today, a situation that few of the major users seem to be aware of. Although there are today other sources for some of ERS products, both within USDA program agencies and in the external public and private sectors, there is clearly no real alternative for much of what ERS can do. If any value is placed on high-quality research, analysis, and secondary data that is independent of political bias or private interest, it is time to act. Although it is difficult to judge, it should be noted as well that some critics of ERS do not believe the agency has used its resources effectively.

In the judgment of one historian of ERS (Bowers, 1990:234-235):

> Its success or failure would depend more than most agencies on the quality of its work, especially as perceived in the Secretary's office. If it could do high quality, unbiased, and timely work, adapt its organization to changing needs, avoid political mine fields, it might prosper. If not, it might meet the fate of its predecessor with no voices raised to defend it outside of the economics profession.

Agriculture Does Not Count Anymore

USDA and ERS have a continuing problem in serving both their older agricultural clientele and all the new clientele that have entered the USDA's legislative coalition since the 1960s. There is a tendency for some groups and individuals in USDA—and outside—to accept an "Agriculture does not count anymore" theory. It is true that the politics of the USDA mission are now dominated by a wide variety of causes and organized interests. And although the political clout of farm and agricultural business groups to create new benefits has been eroded (as has that of most of the older political interests in Washington), they still appear to have the defensive power to protect their interests against assault in the political combat inside USDA and in the Congress.

Consequently, the all-too-common habit of setting one group against another in what is a fragile coalition is politically dangerous, especially for politically neutral units like ERS, which to survive must be allowed to serve the diverse claims on its expertise. As the first director of Agricultural Economics put it, ERS "must be prepared to respond regularly and effectively, without compromising itself, to the economic analytical needs of the Office of the Secretary; it must understand and appreciate the intelligence needs of the Congress and find ways of satisfying those needs without coming into conflict with the administration in power; and it must recognize and anticipate the information and intelligence needs of a diverse national public and develop effective channels for meeting those diverse needs" (Cochrane, 1983:30). It is clear today that ERS cannot achieve this without, at a minimum, both Congress and the secretary's support and understanding of the nature and limits of a federal research and analysis agency. At the same time, ERS must be provided clear political direction with a consistent set of expectations and rules, if its behavior and performance are to meet professional expectations, including those of its diverse clientele.

Impact of ERS Location in USDA

The early BAE (1922-1938), like all USDA agencies, reported directly to the secretary of agriculture. In the 1938-1946 period, during which the BAE became the planning staff for the secretary, the BAE was moved to within the Office of the Secretary. This politicized the BAE and its secondary data and analysis functions and led to the demise of the BAE in 1953, when party control of the executive branch changed.

When the ERS was established in 1961, it reported to a director of agricultural economics (at the assistant secretarial level), who reported to the secretary of agriculture and, for some matters, to the under secretary of agriculture. In that era, an under (later deputy) secretary handled most of the internal day-to-day management of USDA, while the secretary had a full-time job dealing with the Congress, the White House, and the politics of agriculture and clientele groups. The director of agricultural economics (later assistant secretary for economics) was the chief economist and economic adviser to the secretary and an experi-

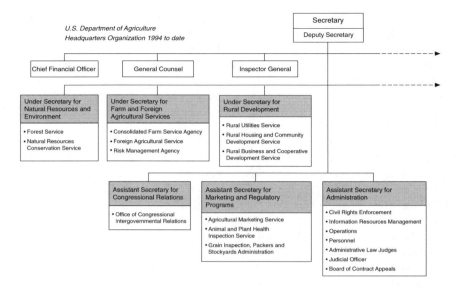

FIGURE 4.2 U.S. Department of Agriculture headquarters organization, 1994-1998.

enced professional economist. Also reporting to the director/assistant secretary was a small policy analysis and advisory group (selected by the director/assistant secretary) that handled highly political, quick turnaround demands, keeping prescriptive political and policy advice at the political level and thus insulating ERS from direct political involvement. This arrangement worked reasonably well from 1961 to 1993. It was less effective in periods when ERS was temporarily merged with two other agencies (1977-1981), and when the Office of the Secretary was held by individuals who operated as their own economist or, more frequently, by individuals who ignored economic analysis and advice and based their decisions entirely on the political process.

In the 1994 reorganization of the Department of Agriculture, all this changed. The assistant secretary for economics was eliminated. Why is not clear. Perhaps the secretary had little use for economic analysis, or perhaps the position was wanted for another purpose (the number of presidential appointment positions in a department is fixed by law); or again, under White House pressure to reduce the number of USDA agencies, internal politics led action agencies and clientele to rid themselves of a highly visible economic adviser with whom they were often at odds. ERS was moved to report to an under secretary for research, education, and economics (see Figure 4.2). The small policy analysis and advisory group, which had reported to the assistant secretary for economics, moved to the Office of the Secretary of Agriculture as the Office of the Chief Economist, where its director became the economic adviser to the secretary.

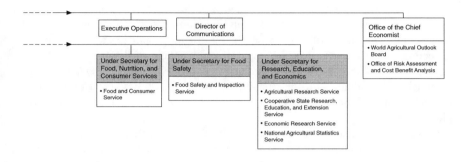

This arrangement has left ERS and NASS isolated from the policy process, making ERS support of the chief economist at best awkward and inefficient. The Office of the Chief Economist is left without the unhindered access to analytic and research support needed to sustain its potential capacity. The current arrangement creates ambiguity in the reporting line for economic policy and the potential to isolate ERS.

ERS Mission

The expectation of the early directors of agricultural economics that ERS would provide all economic work done in USDA was never a realistic goal and was never achieved. Partly in an effort to deal with the analysis that ERS provides the secretary, individual program or action agencies have long employed their own economic researchers and analysts, often hired from ERS ranks. In addition, from the early days of the BAE, the leadership of some action agencies has attempted to acquire the sections of the BAE or ERS and/or the professionals assigned to work on the action agency's programs. Today, more economists work in the action agencies of the USDA than in ERS. The current role of ERS in serving USDA agencies urgently needs to be clarified and stabilized. More attention needs to be devoted by the secretary's office to creating a productive ERS-USDA agency relationship in a USDA environment that is again rapidly changing.

The ERS mission of providing research, shorter-run problem analysis, and secondary data does not appear to have changed. The mix of services has fluctuated with the pressures exerted by the changing problems facing USDA. Although the farm sector, narrowly defined, accounts for less than 2 percent of gross domestic product, the total food and agriculture sector accounts for about 16 percent. The output of this sector is a significant proportion of U.S. exports, reflecting the technology-based comparative advantage of the United States in agricultural products. Besides its food and agriculture mission, USDA has major responsibilities in natural resources and environmental policy, food safety, rural development, and the welfare system, all of which in varying degrees now have global dimensions.

The vastly expanded scope of policy issues and problems needing research and policy analysis has had the most significant impact on the ERS mission. The mandate of USDA, and by extension that of ERS, extends into broad issues of environmental and natural resource policy. The increasing interdependence of domestic economic sectors, the globalization of financial and commodity markets, and the institutional implications of revolutions in information and biotechnology are creating interdependencies that require major multidepartment policy responsibilities. Demand for USDA collaboration in research with other parts of government is growing. This is evident in the diversity of ERS research and information services over the last decade. In part because of the lack of significant intramural economic research capacity in many cabinet agencies, ERS has increasingly been called on for collaborative or contract research.

The demands for new skills and greater diversity in expertise and the large decline over time in real resources and personnel ceilings leave USDA facing the question: Should ERS resources be increased, or should ERS cut the scope of its mission to fit its resources, and if so, where should the cuts come and what is the appropriate balance between different ERS products? It is not clear today what the secretary or the Congress expect of ERS. It is clear that their expectations are frequently inconsistent. This needs to be clarified if ERS is to be effectively supported, organized, and led. ERS cannot by itself resolve this dilemma. It can substantially affect but not control its future.

5

Evaluation as a Framework for Management

The evaluation of government research and information programs in support of public policy can be undertaken at different levels. At the highest level, evaluation involves consideration of the broad objectives of public policy. Consistent with the objectives of this report, our goal here is much more modest. The agenda of the Economic Research Service (ERS), and of economic policy support research agencies generally, derives from the mandate of the economic policy maker to whom the agency reports. In the case of ERS, policy is made by the secretary of agriculture working with Congress.

This chapter concerns the process of evaluating the services provided by ERS. It does not undertake the evaluation itself for any particular service. In fact, it does not even define the units appropriate for evaluation: the specific functions referred to later in the chapter might be too broad and some of the accomplishments too narrow, and in any event definitions of services can be made only by those with day-to-day responsibility for their organization. Instead, this chapter concentrates on a framework for the retrospective evaluation of each service ERS provides. This framework can also be used prospectively, in deciding how each service should be delivered (Chapter 6) and this, in turn, drives the internal organization of ERS and its placement within USDA (Chapter 7). None of the later development is possible without first knowing what the services are and how they will be evaluated.

EVALUATING ERS PROGRAMS

General Guidelines for Program Evaluation

Every day individuals and organizations make decisions and choices. Much of this process involves comparison of goods and services—both those that are available now and those that may become available in the future. The evaluation of a service provided by an organization or individual is the comparison of that service with alternative services that are or could be provided by others. It entails a ranking of the services provided by alternative suppliers. There are no operative absolute standards for providing any service. However, comparison of a service currently provided by a supplier with that provided by a hypothetical supplier can be quite important. (For example, this question may be paramount in contemplating whether to discontinue a service.) Having determined the best provider of a service, the question of whether that service should be offered at all remains open; evaluation is only the first step in this determination.

The evaluation process therefore requires that the service being evaluated be defined, and that alternative providers of the service—perhaps including hypothetical providers—be identified. The relevant comparison of providers of the service is that made by clients—the users and potential users of the service. In an ideal, competitive market for a service, clients compare and rank alternative providers, and each chooses its most favored provider. Food retailing, laundering and dry cleaning, fast food, and a score of other service industries approximate this ideal. Successes and failures of establishments in these industries reflect the evaluations of thousands of clients.

Although a market may reveal evaluations of clients in the aggregate, even a perfect market does not directly indicate the reasons for clients' rankings. Understanding how clients make comparisons is essential to success in a competitive environment. For example, success in the laundering and dry cleaning business requires that managers understand how clients take into account establishment location, hours of operation, promptness of delivery, quality of service, price, and other characteristics in making comparisons and choices. Thus, if evaluation is to be used to improve the delivery of a service, it is necessary to identify those attributes of the service that underlie comparisons by clients.

Many important service markets are far removed from the competitive ideal. There may be only a few suppliers, information may be hard to acquire, and clients may find it very costly to switch suppliers. The same prerequisites for evaluation operate in these situations—namely, it is necessary to identify:

(1) the service provided,
(2) potential providers of the service,
(3) the clients for the service, and
(4) the attributes of the service that underlie clients' comparisons.

To the extent that a market is not competitive, it may indicate less clearly the service, providers, and clients, and the first three prerequisites may be more difficult to satisfy. In any case, it is critical to concentrate on those attributes that are important to the comparisons that clients make.

A specific example of such an imperfectly competitive service market is postgraduate professional education. Professional schools provide services for at least two kinds of clients—their prospective students and those who might hire their graduates. Identification of prospective students is not a straightforward task, for this group is broader than the students who actually apply for admission. For students, alternative providers of the service include, of course, other professional schools in the same discipline, but, if the service is defined more broadly, alternative providers may include professional schools in other disciplines or, even more generally, other career opportunities. Determining the attributes of postgraduate professional education that underlie choices requires study of a fairly wide group of prospective clients. Simply asking students, as they approach graduation, to provide numerical scores for attributes of their professional school is nearly meaningless. Such a survey may uncover comparisons of experience with expectations or vicarious experiences elsewhere, but it does not elicit the attributes that underlie any choice, and comparison of such numerical scores across institutions is of dubious value.

A complete evaluation of an entire program of an organization may be conceived as entries in a four-dimensional matrix of services, potential providers, clients, and service attributes that underlie clients' comparison of alternative providers. Evaluation can be more or less formal. More formal evaluations are more costly than less formal evaluations, but they can provide information of greater strategic value. For example, complaints received by an airline identify attributes of service that are important to clients and are very likely to affect their choice of airlines. But since the complaints are made only by the actual clients of the carrier, the airline is missing an important group of prospective clients—those who are using its competitors' services. A well-designed survey including prospective clients in the sampling frame would be more useful to the airline. As a second example, consider a government agency providing primary data of very high quality to the private sector free of charge. A systematic assessment of clients' comparison of the current data with prospective data of lower quality could provide the appropriate strategic response to a future question of whether to continue the current program, provide lower-quality data at lower cost, or institute user fees to partially or completely cover the additional cost of the higher-quality data.

The concept of evaluation as entries in the four-dimensional matrix of services, providers, clients, and attributes applies to government as well as private-sector programs. When a service is provided by both government and the private sector, there is no important complication if user fees cover the cost of government provision of the service. (Examples involving quasi-governmental organizations include check clearing by the Federal Reserve banks and express package delivery by the U.S. Postal Service.) But this situation is the exception, not the rule. There are very large sectors of the economy—education, transportation, and health care, for example—in which government agencies provide or subsidize services that are also provided by the private sector. This is especially the case in the increasingly important information-based sectors of the economy.

Typically, government provides primary data and some infrastructure, and the private sector provides secondary data and builds on the infrastructure. Leading examples are censuses and large national surveys, including not only the decennial census but also the quinquennial censuses of manufacturing and agriculture and major household surveys, which form the basis for extensive information gathering in the private-sector to support private-sector decision making.

At a conceptual level, the reasons for government involvement in providing information can be traced to the public good arguments set forth in Chapter 2. But at an operational level, determining the demarcation between public and private provision of services requires careful evaluation. Here, the level of user fees for services charged by government agencies complicates the evaluation process. If the private sector also provides the service, then it is important to include price (user fees for the government agency, market price for private suppliers) in the list of attributes. Determining an appropriate user fee for the government service may be difficult, and identifying the hypothetical clientele for this service, if that user fee were charged, is more difficult still. There are two initial steps that can be carried out, however. The first is ascertaining the cost of providing the government service. Since most government agencies jointly produce many services, this may not be straightforward, but essentially the same problem is faced in cost accounting in private firms and methods developed there can be applied in government agencies. A second initial step is to elicit clients' and prospective clients' assessment of their demand at different user fee levels.[1]

The most constructive evaluation processes include hypothetical as well as actual providers of the service. This process is never-ending in any competitive industry and is clearly evident to the casual observer in industries undergoing rapid change. To enter, survive, or thrive, suppliers and potential suppliers are constantly considering and experimenting with new combinations of attributes for services. This is true of successful public-sector markets, for example, the market for basic scientific research in the academic sector and government agencies, as well as in successful private-sector markets. If a government agency is the only provider of a service, then the only alternative suppliers are hypothetical, but the evaluation process is still essential to improving, and even sustaining, the quality of services provided. The principle is taken as granted, even in those cases in which it is given that only government should provide services—most notably, every election involves the comparison by the electorate of hypothetical alternative suppliers of services. It should also be taken as granted in the regular and systematic evaluation of services provided by government agencies, whether or not there are private-sector alternatives for these services.

---

[1]A finding of little or no demand at a reasonable user fee does not immediately imply that the service should be discontinued by the government agency. For example, equity considerations might intervene, as discussed in Chapter 2. But the burden of proof is then on the side of continuing agency provision of the service in question.

Three Types Of Services

In its information and research support of the economic policy function of USDA, ERS provides three kinds of services, staff analysis, secondary data preparation and analysis, and intermediate and long-term research. The most immediate and visible category of service is staff analysis in response to questions from the secretary's office, often from the Office of the Chief Economist. In fiscal 1997, ERS responded to 346 such formal requests. The ERS administrator maintains logs of these requests and the corresponding responses. Table 5.1 provides a list of these requests for a representative period during 1997, and it indicates the diversity of questions to which ERS responds. Responding to these requests takes about 20 percent of the time of the ERS staff of professional employees, most of whom are economists. Table 5.2 provides information about the distribution of ERS professionals by grade level, job title, and division. Table 5.3 provides summary classifications of staff analysis by source of request, assignment to divisions, and timeliness of response for fiscal 1995, 1996, and 1997.

The second category of service is secondary data preparation and analysis. Included in this category are *Situation and Outlook* (*S&O*) and indicator activities. The output of these activities supports policy analysis and is used extensively through subscription by both industry and researchers in analyzing agricultural and food economics. *S&O* and indicator work is most closely associated with commodity market analysis, although indicators are also provided in the areas of natural resources and food economics. The *S&O* work in the commodity area supports the interagency commodity estimates provided by the USDA World Agricultural Outlook Board. (Other USDA agencies participating in the interagency estimate committees are the Foreign Agricultural Service, the Farm Service Agency, and the Agricultural Marketing Service.) Development of indicators and the preparation of *S&O* reports and other information takes about 40 percent of the time of the ERS staff of professional employees.

Much of the market and trade-related *S&O* output is reported through the publication *Agricultural Outlook*, the main source for USDA's farm and food price forecasts. *Agricultural Outlook* emphasizes the short-term outlook for all major areas of the agricultural economy, and also presents long-term analyses of such issues as U.S. agricultural policy, trade forecasts and export-market development, food safety, the environment, and farm financial institutions. The publication presents extensive data on individual commodities, the general economy, U.S. farm trade, farm income, production expenses, input use, prices received and paid by farmers, per capita food consumption, and related issues.

In addition to this general publication, ERS also regularly publishes *S&O* reports several times a year for numerous individual commodities and financial measures (examples include *Agricultural Income & Finance*; *International Agriculture & Trade*; *Cotton and Wool*; *Feed*; *Fruit & Tree Nuts*; *Livestock, Dairy and Poultry*; *Aquaculture*; *Oil Crops*; *Agricultural Exports*; *Rice*; *Sugar & Sweet-*

TABLE 5.1 ERS Staff Analyses: Requests Received during January, 1997

Analysis requested	Received	Due	Sent
Provide background on research, education, and analysis pertaining to global change including responses to specified questions	1/2	1/8	1/8
Assemble a list and cost of subscriptions to newspapers, magazines, other periodicals, and on-line news services	1/2	1/8	1/8
Comment on the thesis that declining rice production in the South has a negative impact on migratory birds and on the environment	1/2	1/10	1/10
Review draft GAO report, "Commodity Programs: Despite Reforms, Some U.S. Prices Will Remain Higher Than World Prices"	1/3	1/8	1/8
Comments and suggestions regarding the coverage content of the Business Expenditures Survey	1/6	1/15	1/17
Provide the number of occupied housing units by region and summarize uses of fuel and uses of electricity	1/6	1/8	1/8
Briefing materials relating to the National Cattlemen's Beef Association	1/7	1/15	1/15
Prepare a 3-part briefing paper relating to the domestic supply, use, and price for dairy over the next year, the international dairy situation re Oceania, EU and GATT commitments, and relation of dairy situation to domestic food assistance	1/7	1/8	1/9
Draft response relating to use of lecithin in Mexican electricity generators	1/7	1/10	1/10
Prepare a briefing paper on the propane price issue	1/8	1/9	1/10
Provide a description of the ERS proposed FY 98 global change program	1/8	1/8	1/8
Briefing materials relating to the Pacific Northwest Flooding	1/9	1/9	1/9
Describe the status of assessments of economic impacts of a worldwide ban on methyl bromide	1/9	1/13	1/14
Review the proposed Western Governors Association memorandum of understanding regarding future management of drought in the west	1/9	1/14	1/16
Provide comments, talking points, and guidance on United Nations Statistical Commission documents	1/9	1/28	1/29
Background on the tax income averaging concept and comments on the Nick Smith bill	1/10	1/17	1/17
Briefing materials relating to Iowa Pork Producers Association	1/10	1/15	1/15
List major Science and Technology related publications issued by ERS over the 1993-1996 period	1/13	1/16	1/16
Discuss the food component of the December Consumer Price Index	1/14	1/14	1/14
Identify hot issues that governors might raise at National Governors Association meeting	1/15	1/17	1/17

continues

TABLE 5.1 Continued

Analysis requested	Received	Due	Sent
Prepare slides of selected charts from briefing handouts and Food Review	1/15	1/28	1/23
Provide information on the civilian labor force, employment, unemployment, and unemployment rate for selected countries	1/14	1/16	1/16
Prepare draft ERS testimony for the appropriation hearings	1/16	1/29	1/28
Update and add/delete Q's and A's for the FY 1998 appropriation hearings ERS witness book	1/17	2/18	2/20
Briefing materials relating to data development associated with the new food safety law	1/22	1/22	1/22
Provide information on persistent poverty counties and Great Plains	1/22	1/31	1/28
Briefing paper on the effect of the January 1997 freeze in Florida vegetable areas	1/23	1/23	1/23
Review and comment on "Working Paper Toward a National Rural Policy"	1/24	1/29	—
Briefing materials relating to California vegetable and fruit growers, including any flood-related issues	1/24	1/29	1/29
Briefing materials relating to small and minority farmers	1/24	2/5	2/5
Provide background information on the Commercial Agricultural Division involvement in the Canada-United States Joint Commission on Grains	1/27	1/30	1/30
Briefing information relating to the Southern Rural Sociological Association and emerging roles of Land Grant Universities	1/27	1/30	1/30
Review and comment on U.S.-Japan science and technology relations and agreement	1/27	2/5	2/3
Briefing materials relating to the winter storms and cold in the Dakotas	1/28	1/29	1/29
Update tables for the 1997 Statistical Abstract of the United States	1/29	3/14	3/19
List of regional centers, consortia, programs, and projects that ERS supports	1/30	2/6	2/6
Review and comment on material for a legislative report on the proposed draft bill on the Treasury Amendment to the Commodity Exchange Act	1/30	1/30	1/31
Review and comment on draft FY 1998 "Our Changing Planet"	1/30	2/6	2/6
Prepare a white paper on industrial hemp	1/31	2/6	2/6
Briefing materials relevant to the National Cotton Council annual meetings in Florida	1/31	2/7	2/6
Review and comment on Summary Report on Class I Price Proposals for Milk Marketing Order Reform	1/31	2/4	2/4

Source: Staff analysis logs provided to panel by ERS.

TABLE 5.2 Social Scientists in ERS by Grade and Job Series (May 23, 1998)

Job Series	Grade Level								
	SES	15	14	13	12	11	9	7	Total
Economist/	2	37	78	137	27	8	4	6	299
Agriculture Economist									
Geographer				1					1
Mathematician					1				1
Operations research analyst			1						1
Social science analyst			1	3	1				5
Social science aid technician								2	2
Sociologist			3	3					6
Statistician		1		1	1				3
Total	2	38	83	145	30	8	4	8	318

Source: Provided to panel by ERS.

TABLE 5.3 Staff Analysis in ERS, 1995-1997

Total Requests	455	456	346
By source[a]			
REE	60 (13%)	93 (20%)	62 (18%)
OCE	125 (27%)	105 (23%)	52 (15%)
Agencies	52 (11%)	58 (13%)	47 (14%)
OSEC	100 (22%)	86 (19%)	106 (31%)
White House	55 (12%)	54 (12%)	26 (8%)
Legislative branch	33 (7%)	22 (5%)	21 (6%)
Other	30 (7%)	38 (8%)	32 (9%)
By division assignment[b]			
RED	151 (23%)	130 (18%)	115 (20%)
FCED	97 (14%)	109 (15%)	94 (17%)
CAD	235 (35%)	281 (39%)	148 (26%)
NRED	113 (17%)	115 (16%)	124 (22%)
OENU	46 (7%)	58 (8%)	63 (11%)
ISD	28 (4%)	28 (4%)	25 (4%)
By timeliness			
On time or early	356 (78%)	378 (83%)	292 (84%)
One day late	60 (13%)	47 (10%)	31 (9%)
Two or more days late	39 (9%)	31 (7%)	23 (7%)

[a]USDA Assistant Secretary for Research, Education, and Economics (REE); USDA Office of the Chief Economist (OCE); USDA agencies (Agencies); Office of the Secretary, USDA (SEC); White House offices including Office of Management and Budget and Council of Economic Advisers (White House); Members of Congress, General Accounting Office, Congressional Budget Office, Congressional Research Service (Legislative branch); remaining clients (Other).
[b]Rural Economics Division (RED); Food and Consumer Economics Division (FCED); Commercial Agriculture Division (CAD); Natural Resources and Environment Division (NRED); Office of Energy and New Uses (OENU); Information Services Division (ISD); Categories sum to more than total requests due to assignments of some requests to more than one division.

ener; *Tobacco*; *Vegetables & Specialties*; and *Wheat*) and a yearbook containing data and related information on an annual basis for most of these commodities. Another important outlet for this type of work is the *Farm Business Economics Report* (formerly called *Economic Indicators of the Farm Sector*), which includes national and state farm income estimates, farm-sector balance sheets, government payments, farm-sector debts, and costs of production by commodity.

ERS produces numerous indicators that summarize the status of natural resource use in agriculture and associated environmental quality. Every few years these indicators are integrated into a comprehensive report, *Agricultural Resources and Environmental Indicators (AREI)*. Following publication of the comprehensive report, as new data and information are collected, ERS publishes *AREI Updates* to supplement and update information contained in *AREI*. *AREI* identifies trends in land, water, and commercial input use, reports on the condition of natural resources used in the agricultural sector, and describes and assesses public policies that affect conservation and environmental quality in agriculture.

Indicators of individual, household, and market food consumption, expenditures, and nutrients, food marketing costs, marketing margins, and farm-to-retail price spreads are also regularly developed and reported by ERS. Periodicals such as *Food Review* and annual publications such as the *Food Marketing Review* and *Food Consumption, Prices, and Expenditures* report data and statistics related to food consumption and nutrition, as well as the structure and performance of the food system.

The third category of service, intermediate and long-term research, accounts for the remaining 40 percent of ERS professional staff time. This research is related to the economic policy mandate of USDA. Currently, this entails a diverse set of projects, as indicated in Box 5.1. This box indicates specific research functions and accomplishments of ERS, organized by division, as detailed by ERS in April 1998. This summary reflects the greatly increased diversity of the policy mandate of USDA, and by implication the diversity of the ERS research program, over the past 20 years, as discussed in Chapters 3 and 4. Table 5.4 provides information on the research publications of ERS professional staff.

The services of data preparation and analysis and intermediate and long-term research, all in support of public policy, are made available not only to public servants charged directly with policy making, but also to private citizens to whom public policy makers are responsible, and who are free to make use of these services in private decision making. As discussed in Chapter 2, in a democracy with a market economy, government might well provide information because it is a pure public good, even though the information was not needed to support the making of public policy. Some ERS services, particularly the provision of reports and indicators, are clearly used for private as well as public decision-making purposes. To determine in any useful sense which ERS services are primarily in support of public policy and which are primarily to provide information as a pure public good is beyond the scope of this report. In any event, in the panel's

BOX 5.1 Specific ERS Research Functions and Accomplishments

Market and Trade Economics Division
Specific function: Conduct research on U.S. and foreign agricultural and trade policies and their relationships to U.S. and world supply, demand, and trade of agricultural products.
1997 accomplishments:
Analysis of China and Taiwan joining the World Trade Organization (WTO);
Support for WTO implementation and future negotiations;
Foreign direct investment and trade;
Technical assistance to emerging market countries.

Specific function: In cooperation with other ERS divisions, analyze the relationships between U.S. food, health and safety, environmental, and rural economic policies and programs and the structure and competitive performance of U.S. and world agricultural markets.
1997 accomplishment:
Study of U.S. agricultural growth and productivity.

Specific function: Develop and maintain an analytic understanding of U.S. and foreign agricultural economic developments, including policy changes and institutional developments that affect agricultural markets.
1997 accomplishment:
Implications of NAFTA for U.S. agriculture.

Food and Rural Economic Division
Specific function: Examine the demographic, social and economic determinants of food and nutrient consumption; interrelationships between food and nonfood consumption; consumer valuation of quality, safety, and nutrition characteristics; and the role of information in determining food choices.
1997 accomplishment:
Estimating nutrition information differentials and their impact on individual diets.

Specific function: Examine the adequacy and effectiveness of government programs, particularly food assistance and nutrition programs, on nutritional adequacy of diets, and food securing including costs and benefits of food assistance and nutrition programs, the extent and social cost of food insecurity, and the role of food assistance in meeting larger goals of welfare programs.

1997 accomplishments:
> Study of low-income household food prices and costs;
> Study of childrens' diets and nutritional shortcomings;
> Evaluating commodity procurement for food assistance programs.

Specific function: Analyze the food processing and distribution sector, including the ability of the sector to meet changing consumer demand; the effect of government market interventions to facilitate that response; and the effect of government interventions and rapid changes in the sector on consumer and producer welfare.

1997 accomplishments:
> Estimating and addressing America's food losses;
> Monitoring and analyzing the U.S. food industry.

Specific function: Analyze food safety issues, including consumer benefits from risk reduction, production trade-offs in reducing hazards, impacts of proposed regulations and international harmonization, and the implications of changing demographics on food safety economics. Also, examine the role played by food safety attitudes, knowledge, and awareness in shaping food choices and eating behavior.

1997 accomplishments:
> Economic assessment of the new meat and poultry inspection system;
> Benefits of improved drinking water quality;
> Estimates of societal costs from *Campylobacter*-associated Guillain-Barré syndrome.

Specific function: Analyze the economic, social, and demographic factors influencing the infrastructure of rural communities, agribusiness activity, and the industrial base of rural areas. In particular, analyze the development of rural portions of geographic regions of the United States, including changes in industry mix, tax policy, credit availability, and other economic activities, and means of measuring overall economic development.

1997 accomplishments:
> Rural credit study;
> Rural empowerment zones;
> Comparing income and wealth of farm operator households with all U.S. households.

Specific function: Determine the effects of economic, social, and governmental policy behavior on the demand for and supply of state and local government services including low-income

Box 5.1 Continued

assistance programs, the quality of such services, and the
relationships between local services and the viability of families
and communities, particularly in rural areas.
1997 accomplishment:
   Estimating the effect of electric utility deregulation on rural
   communities.

Resource Economics Division
Specific function: Evaluate the implications of resource
conservation and environmental policies and programs on
commodity prices, producer and consumer welfare,
competitiveness, and sustainability of land and water resources.
1997 accomplishments:
   Documenting agricultural resource and environmental trends;
   Evaluating the benefits, costs, and role of conservation tillage;
   Assessing the use of partial interests in conservation policy.

Specific function: Analyze the impacts of global change (climate
and other resource adjustments) and international policies on
production, competitiveness, and environmental quality.
1997 accomplishment:
   Analyzing climate change effects and policies.

Specific function: Assess the economic and environmental effects
of resource-conserving production management systems.
1997 accomplishment:
   Support for Conservation Reserve Program implementation.

Specific function: Analyze the costs, benefits, and distributional
impacts of technologies designed to reduce environmental risk
associated with agriculture.
1997 accomplishments:
   Improving understanding of integrated pest management;
   Surfacing viewpoints on the economics of agricultural
   sustainability.

Specific function: Assess expenditures, returns, and comparative
advantages of public and private research funding.
1997 accomplishment:
   Confirming international benefits of agricultural R&D.

Specific function: Assess structural change in agriculture, including
the factors affecting structural change and the implications for
agricultural, technology and resource policy.

1997 accomplishments:
> Vertical coordination in the U.S. port industry;
> Foreign direct investment and trade;
> Analyzing the extent of change in the structure of U.S.
>   livestockfarms.

Specific function: Estimate how farm sector financial performance is related to changes in farm policy (commodity, technology, and resource) and financial viability.
1997 accomplishments:
> Analysis of producer risk issues;
> Contracting as a business option for a growing share of
>   farmers;
> Planting flexibility and price volatility under the 1996 farm act;
> Value-added estimates for the U.S. farm sector.

Office of Energy
Specific function: Conduct a program of economic analysis on energy and energy-related policies and programs.
1997 accomplishments:
> Estimating the effect of elimination of the federal ethanol tax
>   exemption on agriculture;
> Review of Department of Energy greenhouse gases and corn
>   ethanol report.

Specific function: Conduct a program of research on the feasibility of new uses of agricultural products. Assist agricultural researchers by evaluating the economic and market potential of new agricultural products and techniques in the initial phases of development and contributing to prioritization of departmental research agendas.
1997 accomplishments:
> Biodiesel feedstocks report;
> Estimating ethanol cost of production for small plants;
> Life-cycle analysis of biodiesel versus petrodiesel.

Source: From ERS website. Specific functions located at:
www.econ.ag.gov/AboutERS/
1997 accomplishments located at:
www.econ.ag.gov/AboutERS/accomp97.htm

TABLE 5.4 Research Publications of ERS Professional Staff

Staff Publications in Refereed Academic Journals, 1992-1996

Year	Number of Articles	Solely Authored	Jointly Authored
1992	156	58	98
1993	147	48	99
1994	113	42	71
1995	113	36	77
1996	92	24	68
Total	621	208	413

Most Common Journals, 1992-1996

Journal	Number of Articles
American Journal of Agricultural Economics	56
Journal of Agricultural Economics Research	29
Canadian Journal of Agricultural Economics	15
Agribusiness	14
Land Economics	14
Agricultural Economics	6
Policy Studies Journal	6

Internal ERS Reports, 1994-1997

Subject Area	Number of Reports
Agriculture and food policy	5
Agricultural research and development	3
Banking and farm credit	9
Country and regional topics	28
Farm programs	3
Farm sector economics (prices, financial conditions, income, structure)	13
Farmworkers and farm employment	4
Field crops	40
Food (advertising, assistance programs, consumption, costs, marketing, safety, spending)	35
Inputs and technology (biotechnology, fertilizer, pesticides, climate)	22
Land, water, and conservation	22
Livestock, dairy, and poultry	20
Rural development, income, and employment	20
Specialty agriculture (e.g., aquaculture, potatoes)	26
Trade issues (food aid, GATT, NAFTA)	20
U.S. agricultural trade	6
Total	276

Source: Provided to panel by ERS.

judgment, there are no ERS services that clearly are not provided in part to support public policy making. As noted in Chapter 4, however, ERS services with a substantial private-sector clientele, including many of the reports and indicators, have been a source of political support for the agency.

Current Framework for Program Evaluation

The most important characteristic of the ERS mission, plans, and goals is that, taken together, they provide the foundation for evaluation of the ERS program. As mandated by the Government Performance and Results Act of 1996, USDA has prepared a strategic plan for 1997-2002 (U.S. Department of Agriculture, 1997). This plan includes the following mission statement for ERS (p. 7-61):

> The Economic Research Service provides economic analysis on efficiency, efficacy, and equity issues related to agriculture, food, the environment, and rural development to improve public and private decision making.

The ERS strategic plan identifies five goals (pp. 7-61–7-65):

> The agricultural production system is highly competitive in the global economy;
> The food system is safe and secure;
> The Nation's population is healthy and well-nourished;
> Agriculture and the environment are in harmony;
> Enhanced economic opportunity and quality of life for rural Americans.

In support of each goal is an objective. For example, the objective in support of the first goal is (p. 7-61):

> Provide economic analyses to policy makers, regulators, program managers, and those shaping public debate that help ensure that the U.S. food and agriculture sector effectively adapts to changing market structure, domestic policy reforms, and post-GATT and post-NAFTA trade conditions.

For each objective, there is a statement of strategies for achieving the objective. The strategy statement for the first objective is (p. 7-62):

> Identify key economic issues relating to the competitiveness of U.S. agriculture, use sound analytical techniques to understand the immediate and broader economic and social consequences of alternative policies and programs and changing macroeconomic and market conditions on U.S. competitiveness, and effectively communicate research results to policy makers, program managers, and those shaping the public debated regarding U.S. agricultural competitiveness.

Finally, there are performance measures corresponding to each goal. For the first goal, the performance measures are "Reports, briefings, staff papers, articles, and responses to requests that provide . . ." followed by a list of seven substantive topics, for example, "economic analyses on the linkage between domestic and global food and commodity markets and the implications of alternative domestic policies and programs for competitiveness."

The objectives, strategies, and performance measures corresponding to the other four goals are quite similar.

With respect to evaluation, the ERS strategic plan states in its section on "linkage of goals to annual performance plan" (p. 7-67):

> Performance measures will assess the extent to which policy makers, regulators, program managers, and organizations (including major media) affecting the public policy debate have high-quality, comprehensive, objective, relevant, and accessible economic analyses for senior policy officials. . . . ERS will use metrics to partially describe its volume of output. . . . The annual performance reports also will include narratives covering characteristics of ERS output that demonstrate that ERS analyses were high quality, objective, relevant, timely, and accessible. The narratives will cover ERS anticipation of issues and the timeliness of output, review prior to release, customer views on relevance and accessibility of ERS analyses, and how ERS analyses contributed to informed decision making.

Evaluating ERS Services

The ERS mission statement and the goals, objectives, and performance measures of the ERS strategic plan concentrate on the substance of ERS research and information provision. The performance measures identify services at levels approaching, but still broader than, the level required for the four dimensions of evaluation described earlier in this chapter. Assessment of specific services by specific clients is raised only in the linkage of goals to the annual performance plan. This section of the strategic plan also identifies five global attributes for ERS analyses: quality, objectivity, relevance, timeliness, and accessibility.

With respect to the framework for evaluation discussed above, however, the ERS strategic plan has three critical shortcomings. First and most important, it gives no indication of comparisons of ERS services with either existing or hypothetical alternatives. It does not hint of comparing ERS performance with that of any other organization along any quantitative or qualitative dimensions. Second, the strategic plan makes no provision for assessing the costs of providing any specific service, or evaluating costs in the light of the quality or other attributes oã the service provided. Third, the plan concentrates largely on setting forth the substance of what ERS currently does, at the expense of focusing on improving the relevant attributes of the services it delivers.

RECOMMENDATION 5-1. Taken together, the ERS mission statement, strategic plan, and annual performance plan should identify the services provided by ERS, the clients and potential clients for each service, potential providers for each service, and the attributes of each service critical

to evaluation. An effective system of program evaluation will seek to establish the competitive position of ERS with respect to the services it provides and the reasons for that position.

Mission

To be effective as the driving force for an organization, a mission statement must explain the function of the organization with respect to the services provided and prospective clients. The ERS mission statement should be sufficiently broad that it rarely needs modification, whereas the strategic plan for a branch for one year should be much more specific and change from year to year. The current ERS mission statement defines one very broad product (economic analysis) and indicates the broad substantive scope of this analysis. This product is further specified in the specific functions of each ERS division, presented in Box 5.1. In fact, the substantive scope of ERS work derives from the mandate of USDA and, as detailed in Chapters 3 and 4, this derived mandate has greatly changed throughout the history of the BAE and ERS, especially in the past 20 years. In fact, as indicated earlier in this chapter, ERS provides secondary data as well as economic analysis. The mission statement of ERS should recognize its functions and indicate broadly how those functions are carried out.

RECOMMENDATION 5-2. The mission of ERS should be to provide timely, relevant, and credible information and research of high quality to inform economic policy decision making in USDA, the executive and legislative branches of the federal government, and the private and public sectors generally. It should identify information and frame research questions that will enhance and improve economic policy decisions within the authority of the secretary of agriculture, organize the subsequent collection of information and conduct of research, and evaluate alternative approaches to policy problems. The work of ERS should address anticipated as well as current and continuing policy questions.

Services

Services provided by ERS should be defined both narrowly enough that ranking them against alternative potential suppliers is possible and broadly enough that a workable group of clients for the service can be identified. For purposes of organization, it is natural to group substantively related services into branches and divisions within the agency, but it makes little sense to try to compare how ERS analyses "help ensure that the U.S. food and agriculture sector effectively adapts to changing market structure, domestic policy reforms, and post-GATT and post-NAFTA trade conditions" along the attributes of quality, objectivity, relevance, timeliness, and accessibility, with other actual or hypothetical providers, because the topic is so broad. To do the same with "economic analyses on the linkage between domestic and global food and commodity markets and the implications of alternative domestic policies and programs for competitiveness," is more realistic, but it remains a task requiring further organization. If one moves to the level of 5- and 10- year projections of agricultural commodity prices, the job is manageable. For a given substantive activity of ERS—for example, economic analyses on the linkage between domestic and global food and commodity markets and the implications of alternative domestic policies and programs for competitiveness—it may be important to distinguish between intermediate and long-term research, monitoring, reporting, and staff assignments as separate services.

Clients

Clients are the potential users of the identified service, whether that service is provided by ERS, a private firm, an international organization, or another government agency. Clients, and especially potential clients, may be difficult to identify. The universe of "policy makers, regulators, program managers, and organizations shaping public debate of economic issues," so frequently mentioned in the ERS strategic plan, is a large one. An essential task of ERS management is to identify the subset of this important group of individuals and organizations for each service it provides or contemplates providing. In the case of 5- and 10- year projections of agricultural commodity prices, the primary client has been the USDA Budget and Program Analysis Office, because the impact of these programs on future federal budgets depends substantially on future prices. As mandated in the farm bill of 1996, these support programs are gradually being reduced and may even be eliminated, in which case for these purposes the Budget and Program Analysis Office will no longer be a client for these services. There are other clients, for example, the Office of Management and Budget and the Congressional Budget Office. Determination of clients and potential clients for this particular service is essential to evaluating the service, and ultimately in deciding whether ERS should continue to perform this function.

Providers

Once services are identified at an appropriately narrow level, other actual providers of the service should not be hard to identify. In the case of information and research services, this amounts to knowing the secondary sources. Identifying potential providers of the service is a more sophisticated, but reasonable, task. As elaborated in the next chapter, this task is an essential first step in deciding whether a particular information or research service should be provided directly by ERS or should be procured from another party. In the case of 5- and 10-year projections of agricultural commodity prices, another provider is the Food and

Agricultural Policy Research Institute (FAPRI) based at the University of Missouri and Iowa State University. Neither ERS nor FAPRI charges user fees for these projections; both receive congressional appropriations. Whether alternative providers would emerge and the relevant attributes of the projections they would provide, if ERS and FAPRI were either to charge user fees or to cease providing projections, is the sort of comparison with a hypothetical provider that is essential to serious evaluation.

Attributes

Simply eliciting clients' comparisons of the providers of an identified service is not enough. If ERS is to improve its delivery of services on the basis of evaluation, it must know why clients rank providers as they do. This is no less true if the ERS service is the hands-down favorite among clients than it is if potential clients universally disdain the ERS service. In many cases, ERS managers will have a good indication of the important broad attributes—for example, quality, objectivity, relevance, timeliness, and accessibility—but it is critical to obtain clients' open-ended assessment of the reasons for the comparisons they make. For example, in the case of 5- and 10-year projections of agricultural commodity prices, the client might indicate that the FAPRI provides projections earlier, but only by regions of the world, whereas ERS provides projections country by country.

When a government agency does not charge user fees for its services, it is easy to overlook the cost of the service in question as one of its key attributes. Currently ERS makes no provision for allocating costs or staff time among the services it provides. It is not difficult to record this information—indeed, for ERS operations supported by interagency transfers (for example, from the Agency for International Development), this sort of accounting is maintained. Without accounting for costs and staff time according to the services provided, any evaluation tool will be of quite limited usefulness.

RECOMMENDATION 5-3. ERS should allocate its costs and staff time across the services used in its system of evaluation, according to generally accepted accounting principles.

Evaluation Process

The evaluation process can be either formal or informal. Regular informal contact between providers and clients in a competitive environment leads to ongoing evaluation, for which any formal process is likely to be a poor substitute. This model is very familiar in private markets, including those for information and research embodied in products like newsletters and consulting. It is also familiar in the public sector, in the form of expert advisers (for example, the chairman and members of the Council of Economic Advisers) who are appointed to serve their primary client (the president, in this example). For ERS programs, this model is not directly applicable, and a more formal evaluation process is required. More formal approaches are also mandated by the Government Performance and Results Act. Formal instruments for evaluation should be organized along the four dimensions described here. In addition, the evaluation should be administered by a third party, and, ideally, clients and potential clients should not know that ERS is sponsoring the evaluation. Examples of third parties include the Measurement Laboratory of the Bureau of Labor Statistics and private-sector auditing and consulting firms.

RECOMMENDATION 5-4. Formal program evaluation instruments should elicit from clients and potential clients their choices among alternative providers and potential providers of the services provided by ERS, and the attributes of the services critical to their choices, including prices. The instrument should solicit the identities of additional potential clients and alternative providers of these services. ERS should participate in the design of evaluation instruments, but their administration should be delegated to an independent party.

EVALUATING INDIVIDUALS

The evaluation of professionals in ERS or any other organization poses a set of questions distinct from the evaluation of programs. Within ERS, the dimensions of evaluation for economists are set by the Office of Personnel Management (OPM). The five dimensions are scope of assignment, technical complexity, technical responsibility, administrative responsibility, and policy responsibility. Some examples from current OPM policy (U.S. Office of Personnel Management, 1996) indicate the kinds of qualities involved.

- With regard to scope: "The GS-13 economist must initiate, formulate, plan, execute, coordinate, and bring studies to meaningful conclusions."
- With regard to technical complexity: "GS-14 economists are almost entirely dependent on their own personal professional knowledge and imagination in the assessment and understanding of problems of critical importance."
- With regard to technical responsibility: "The GS-12 economist is accountable not only for the factual accuracy of his results but for the thoroughness of his research plan and the cogency of his interpretations."
- With regard to administrative responsibility: "Subject to supervisory approval, economists at the GS-13 level are responsible for identifying, defining, and selecting specific problems for study and for determining the most fruitful investigations to undertake."

• With regard to policy responsibility: "GS-14 economists serve as authoritative technical advisors, within the area of assignment, in the highest councils of Government."

In 1996, ERS initiated an Economist Position Classification System, based on the OPM economist standard. Under this system, "economists have open-ended promotion potential based on their personal research and leadership accomplishments, which can change the complexity and responsibility of their positions" (Economic Research Service, 1997:1). Evaluation is carried out by a peer review panel, whose chair has final authority for determination of position grade level. This system supplements, but does not replace, annual reviews by immediate supervisors. Peer panel reviews are conducted every three years for positions GS-12 and below, every four years for GS-13, and every five years for GS-14 and above. There are provisions for early and delayed reviews and for reevaluation. The objective of the peer panel review is to grade the incumbent against the five dimensions of the economist classification standard and assign the grade level that best matches the incumbent's qualifications.

General Guidelines for Evaluation of Individuals

The principle of comparison is as valid for the evaluation of individuals as it is for the evaluation of programs. The objective is to rank an individual's performance relative to the performance of other individuals in similar positions doing similar work. There can be no absolute standards. Regardless of the kind of work being done, any aspect of individual performance must satisfy five characteristics appropriate for evaluation, discussed below. The way that evaluation is carried out differs greatly, depending on the universe of comparison for individual performance.

Characteristics of Individual Performance

An aspect of individual performance is *consequential* to the extent that it directly affects the attributes of services provided by ERS that are identified in the program evaluation process set forth earlier in this chapter. An individual aspect of performance is *controllable* if the individual has substantial control over that aspect of his or her work. Ascertaining those aspects of performance that are controllable is more difficult for individuals working in teams than it is for individuals working alone. An aspect is *observable* if the individual's immediate supervisor can monitor that aspect of the individual's performance. It is *verifiable* if the supervisor's observation can be replicated by others. Organizing tasks so that consequential aspects of individual performance are also observable and verifiable is a key task of management in any organization. A good organization of tasks will provide individual incentives that support favorable program evaluations; if tasks are poorly organized, then it is not possible to reward individuals for performance that supports program objectives. Finally, an aspect of performance is *comparable* if it exists and can be measured in much the same way as the performance of others. This characteristic presents special problems, to which we return shortly. First, we take up three examples that illustrate how these characteristics are important to the evaluation of individual performance.

Consider an individual who is responsible, in part of his or her position, for written staff assignments in response to requests for information coming directly from the Office of the Secretary. An important aspect of the individual's performance is whether or not the written response addresses the question posed. This is largely controllable by the individual, although not entirely—for example, time constraints and the availability of data or other pertinent information must be taken into account in trying to isolate how well the individual addressed a particular question. (As this activity is repeated, these uncontrollable characteristics may be about the same for this individual and for those with whom he or she is compared.) This aspect is also an important attribute in the evaluation of the staff analysis services of ERS. If the individual's supervisor is appropriately qualified, then this characteristic is observable, and it can be verified by asking the ultimate client, the secretary, whether the response was relevant—indeed, this information is likely to be volunteered if the information provided is badly off target. Since there are many professionals performing similar activities in ERS and other agencies, this aspect of the individual's performance is comparable.

In the second case, consider a senior economist with substantial discretion and responsibility for intermediate and long-term research in support of a service provided by ERS: for example, economic analysis of alternative designs for the auction process used in the Conservation Reserve Program, or the synthesis and commissioning of studies on the impact of agricultural policy changes on carbon dioxide emissions. The related academic publication and citation record of the senior economist, or for the studies that he or she has managed, is consequential to the quality and credibility of ERS research, as well as controllable, observable, verifiable, and comparable.

In the final example, consider an individual who has had discretionary responsibility for a computational general equilibrium (CGE) model, used to address questions about the impact of changes in trade agreements and tax policy. The quality of the model directly affects the quality, and ultimately the credibility, of ERS responses to questions about trade and tax policy—both important attributes in the evaluation of this service of ERS. Although the model is important, it uses established and proven methods, and its use is therefore not a candidate for publication in refereed journals. In both program and individual evaluation, critical anonymous reviews of the model might be solicited. Direct or indirect comparison of this individual with others might be complicated by the unique characteristics of the problems addressed by the CGE model and the technical complexity of the work.

RECOMMENDATION 5-5. The evaluation of professional staff should be grounded in aspects of individual performance that directly affect attributes of ERS services. The aspects of an individual's performance that are evaluated should be under the individual's control and should be capable of being verified by ERS staff beyond the individual's immediate supervisor.

The three examples taken up illustrate that the appropriate method of evaluation differs greatly depending on the kind of activity in which the individual is engaged. The OPM grade definitions are based on the complexity of the tasks assigned to a position. The Economist Position Classification System currently implemented by ERS is a method for assessing these complexities in some depth. Neither, however, directly addresses the fundamental question of how successful the individual in the position is in carrying out the tasks important to the mission of ERS, relative to the others in similar positions. Doing so requires greater flexibility and imagination than has so far been achieved in federal agencies.

RECOMMENDATION 5-6. Standards for evaluation of professional staff should be driven by the tasks that are important to the success of ERS programs. Standards of evaluation should therefore be different for professional staff engaged in different activities. No one standard is appropriate for all economists, much less for all professional staff.

RECOMMENDATION 5-7. Career evaluation of professional staff should be conducted by supervisors and appropriate peers, including ones outside ERS. In each case, it is essential that these evaluators gather information from the widest appropriate sources. Sources include clients for ERS services, external critical evaluators of technical work retained for the purpose, and publications and citations of research.

The ease with which individuals can be compared with their peers, and the universe of peers, varies greatly across positions. For example, many research assistants perform similar functions, and most senior economists have worked with quite a few research assistants in their careers. Comparing the work of research assistants is therefore a more or less straightforward matter. Very senior economists engaged in intermediate and long-term research do work that is typically impossible for supervisors to observe directly, due to its technical specificity and complexity, but the institutions of peer-reviewed academic journals and journal article citation provide systematic comparisons by experts, with observable and verifiable outcomes.

For other positions, comparison by any means can be quite difficult. One example is provided by information specialists who have responsibility for monitoring and reporting in particular substantive areas. Consider a specialist in ERS with sole responsibility for monitoring and reporting for a crop or group of crops. If there is no other specialist with that responsibility anywhere—including other government agencies as well as the academic and private sectors—then comparisons with actual alternatives cannot be elicited in either the program or individual evaluation exercises. Comparison with hypothetical alternatives is problematic, and if comparisons are made with specialists in other areas, then sorting out how much of the difference is controllable is likely to be difficult.

As a second case, return to the hypothetical example previously set forth in which there is no market for comparison of the CGE model used to address trade and tax policy changes, as there is for academically innovative research papers. Since the work is technical and intricate, the details are likely to be observable and verifiable only at high cost, and determining the extent to which the quality of the outputs is controllable is nearly impossible. More generally, the issue of how to deal with positions in which individuals control key aspects of technical processes is a universal and increasingly difficult and important one in both private- and public-sector organizations.

It is essential for managers to prevent positions involving critical, unique, and complex tasks from becoming bottlenecks to the success of agency programs. To this end, ERS should make openness and transparency in performance of these functions central in the evaluation of individuals in those positions. One device for doing so is to require that such positions include documentation of the work performed, to the level that a new individual could assume the position with minimum disruption to the services provided by ERS. The evaluation of individuals will then depend in large part on the quality of this documentation, which in turn would be evaluated by those who would potentially replace the individual if performance were substandard.

CONCLUSION

Evaluation of ERS programs must be conducted against world standards, not simply against the best that can be produced within ERS as a strictly intramural research and information agency. This is the appropriate standard for research and information in support of public policy in the United States, and it is especially important for ERS at this time, as its mandate is extended well beyond the training and background of much of its permanent staff of professionals. An open system of evaluation can do much to ensure that perceptions of the quality of ERS work are consistent with reality. A continuous, systematic program of evaluation will also help to insulate ERS and USDA from shifts in course as division directors come and go. Systematic evaluation of individuals on program-related criteria will facilitate internal rewards to good work.

A well-articulated system of evaluation of ERS programs is the appropriate cornerstone for management decisions within ERS. An effective and regular system of program evaluation will provide information essential to determining the appropriate scope of ERS activities, to evaluating the performance of individual professional employees, to management and allocation of programs among potential suppliers, and to meeting ERS responsibilities under the Government Performance and Results Act of 1996. The next two chapters establish the line between the evaluation process and the internal administration and organization of ERS, respectively.

6

Administration of Research, Information, and Policy Analysis

The management and organization of research, information, and policy analysis in the Economic Research Service (ERS)—or any policy support research agency—should be designed to obtain the best possible outcome in evaluating the program of the agency. This chapter endeavors to carry through this exercise in its broadest outline. It begins by outlining the services that ERS is called on to perform to carry out its mission and the way these services and their clientele have changed rapidly in recent years. At the same time, the universe of potential suppliers of the services ERS provides has greatly expanded, and several alternatives to a strictly intramural research and information program have emerged. The chapter then goes on to examine what are likely to be important attributes of most of the research, information, and policy analysis services provided by ERS and the prospects for favorable evaluation in these dimensions under different models of administration. It applies the principles of program evaluation, developed in the previous chapter, to the problem of delivering research, information, and analysis in support of economic policy making in the U.S. Department of Agriculture (USDA).

This chapter emphasizes how research, information, and policy analysis should be administered in ERS, leaving open questions of who should be doing what. How ERS should be organized, and where it best fits within USDA, are taken up in Chapter 7.

CHANGING MARKETS FOR RESEARCH AND INFORMATION

The substantive mission of ERS derives from the economic policy responsibilities of USDA. This has been true since the organization of its predecessor agency, the Bureau of Agricultural Economics (BAE), in 1922. In the intervening years, USDA's policy mandate has become increasingly diverse and complex, and in consequence so have the research and information programs in ERS. Today, there are many institutions and individuals providing research and information similar to that provided by ERS, and the organization of these activities takes a variety of forms.

Increasingly Diverse Responsibilities and Providers

At its inception, the BAE concentrated almost entirely on production agriculture, with a dominant focus on understanding farm firms and the farm sector as a subset of the economy. By the time of its demise in 1953, local and regional microeconomic issues related to farm income and rural welfare had been added. Marketing research, labor economics, and welfare issues were added in the 1950s, and the analysis of foreign agricultural markets, environmental quality, nutrition, and rural economic development in the 1960s. During the 1970s and 1980s, ERS took on additional tasks in support of trade negotiations. Despite the recent move away from central control and toward freer markets in agriculture, multiple policy goals continue to generate issues that demand new information and economic analysis in support of decision making. These demands are well reflected in the list of specific ERS functions and accomplishments presented in Box 5.1.

In recent decades, the number of ERS personnel has declined, having reached its peak in the 1970s and early 1980s. At that time, the production of indicators alone accounted for about a third of the ERS budget. Often two or three professionals, and in the case of major commodities many more, were assigned to the same market. The responsibilities of these individuals included research, economic modeling, market organization, data base design, and the production of secondary data. Today one individual is assigned to each market, with responsibilities limited to indicators, *Situation and Outlook* reports, and occasional help with staff analyses. An increasingly large share of ERS staff is responsible for research and staff analyses on the wide variety of topics documented in Box 5.1.

At the time of the inception of the BAE, some agricultural economics departments had been in existence since the early part of the century, but many were only beginning to be established in the land grant colleges and universities. A few early institutions produced most of the BAE economists; it was about the time of World War II or afterward before the total number of agricultural economists on college and university faculties exceeded the number of agricultural economists in the BAE. Today, several federal agencies, including the National Science Foundation and the National Institutes of Health, have extramural research programs in economics. Government-sponsored economic research takes place in a wide variety of institutions, including universities, government agencies, federally funded research and development centers, wholly funded academic research institutes (for example, the Food and Agricultural Policy Research Institute centered at the University of Missouri and Iowa State University, and the Institute for Research on Poverty at the University of Chicago, Northwestern University, and the University of Wisconsin), private foundations, and for-profit private consulting firms. Seventy-five years ago, the BAE was the dominant employer of agricultural economists, and the intramural agency research program was the only model for policy support research. Today, the ERS research program remains almost entirely intramural, but it is one employer of agricultural economists among many, and there are several alternatives to the intramural agency research model. Whereas the BAE was, in its early days, the dominant employer of agricultural economists and virtually the sole supplier of economic information and research in agriculture, today ERS must compete with others in hiring professionals and in the provision of many kinds of research and information—and do so across a much wider range of professions and problems.

Some Alternative Models for Economic Policy Support

Other models exist in government, in the United States and elsewhere, for organizing research in support of policy analysis. The following three examples include a special operating agency within government, a nonprofit research institute outside government, and an internal research management agency relying primarily on outside expertise.

Australian Bureau of Agricultural and Resource Economics

Over the past decade, Australia has changed the structure of its government-sponsored agricultural economics research. The Australian Bureau of Agricultural and Resource Economics (ABARE) was formed in 1987 when two much older units, the Department of Primary Industries (DPI) and Department of Resources and Energy (DRE), were combined. As part of this merger, the Bureau of Agricultural Economics from DPI was combined with the Bureau of Resource Economics from DRE to form ABARE. ABARE's activities are similar to those of ERS. It provides a combination of intermediate and long-term research, short-term analysis for policy decisions, and data collection and development of indicators.

Reporting lines for ABARE differ from those of ERS. ABARE is professionally independent, but its executive director reports directly to the minister of primary industries and the minister of resources, who have broad mandates addressed by ABARE research. In addition to reports and other publications that are distributed widely, ABARE also provides briefings to the ministers on issues addressed by its research program.

About 60 percent of the budget of ABARE is provided as a direct appropriation from the Australian parliament. As a professionally independent entity, its research reports are not subject to clearance by any Australian government agency. The balance of its budget comes from a combination of private-sector, regional, and local government units and other federal ministries. Its research priorities are driven by funding sources. ABARE is gradually breaking the ties between its staff salaries and those of the Australian government.

ABARE research reports are well regarded and generally meet high professional standards. Its annual reports document several score of research programs, and it tracks performance indicators for each of them. However, its research independence from the economic interests of some of its private sponsors has recently been questioned, particularly its reports on global climate change, which were funded by Australian energy interests. Some of ABARE's analyses of trade policy have been criticized as too simplistic.

Although ABARE provides economic policy analysis for agriculture, its operating environment differs sharply from that of ERS. In particular, the Australian economy depends heavily on exports of agricultural commodities and natural resources. Australian policy is intensely free trade, with weaker competing interests for this policy than is the case in the United States. Furthermore, since the Australian government is parliamentary in form, ABARE is not subject to different policy positions taken in the executive and legislative branches.

Food and Agricultural Policy Research Institute

The Food and Agricultural Policy Research Institute (FAPRI) at the University of Missouri and Iowa State University is a nonprofit research institute established in 1984. It is affiliated with five other universities as well, drawing on a broad range of expertise from these institutions to conduct a research program that centers on the production and consumption of agricultural products, product prices, farm income, financial risk and risk management, and foreign agricultural trade and policies. (All of these research topics are addressed by ERS as well.) FAPRI receives a direct annual appropriation from Congress and some state legislatures, and it is free to raise its own funds from the private sector, foundations, and foreign governments. It is not considered a federal agency and its staff are not constrained by Office of Personnel Management (OPM) rules.

Much of FAPRI's work is based on its commodity and trade econometric models, together with its industry expertise, in which it clearly has a long-term interest. FAPRI also generates secondary data—both indicators and reports— similar to those provided by ERS. It responds to congressional inquiries with staff analyses in much the same way that ERS reports to the secretary of agriculture. The development of a long-term, independent program of research at FAPRI is hampered by the year-to-year nature of its core congressional funding. FAPRI has been able to devote resources to the long-term refinement of the models it uses in policy analysis. In particular, FAPRI's research, other than model development, has tended to be in response to policy questions well along in the decision process.

Office of Program Analysis and Evaluation

The Office of Program Analysis and Evaluation (PAE) in the U.S. Department of Defense (DOD) has a broad mandate for research and analysis related to the DOD budget. Some of the topics it has addressed include health care for military and dependent personnel, schooling of dependent children, state-by-state impacts of DOD budget changes, and development and maintenance of a model for retirement of military personnel. Professional staff in PAE prepare staff analyses for the DOD secretary, frame questions for intermediate and long-term research, and write and administer research contracts and requests for proposals. There is almost no intramural research.

PAE contracts primarily with federally funded research and development centers (FFRDCs), but also with private-sector research firms. It has long-term relationships with several FFRDCs, and these organizations have in turn made long-term investments in models that are used to support PAE staff analyses. Although the primary relationship with these external vendors involves the development and completion of multiyear research contracts, FFRDCs are also able to respond to occasional specific research and modeling requests within one month. In addition, they have been used to conduct highly classified research. Involvement in political decision making is entirely the responsibility of internal PAE staff.

The FFRDCs utilized by PAE have developed long-term relationships with several federal departments, including Labor and Health and Human Services as well as DOD. Relative to federal agencies and universities, FFRDCs are able to marshal concentrations of very talented professionals for specific assignments, in large part because they are not subject to Office of Personnel Management rules or academic tenure agreements. At the same time, their volume of research and long-term relationships enable them to make multiyear commitments to their core professional staff.

The critical component of the PAE model is its small, internal staff of professionals. These individuals must understand the practical nature of policy questions and frame the relevant questions for economic research. They must also develop and nurture long-term relationships with a variety of vendors, with due attention to incentives for their contractors to perform research that is relevant and meets high standards.

CRITICAL ATTRIBUTES OF ERS SERVICES

The programs and staff of any agency should be managed for the best possible outcome. If there are well-articulated standards for evaluation, then it is possible to work prospectively from these standards to the appropriate administration of research, information, and policy analysis. Of the four dimensions of evaluation described in Chapter 5, the most important one for this process is the attributes of the services that ERS provides. Although specifics of these attributes will vary with the service provided, there are four that drive the agency-wide administration of programs: (1) timeliness—delivering research, information, and staff analysis when it is most effective or required; (2) relevance—providing research and information that assists decision making; (3) quality—meeting the analytical standards of the relevant scientific disciplines and applied professions; and (4) credibility—ensuring that research is and is perceived to be dispassionate and independent of political influence. None of these attributes stands alone. All are likely to interact in clients' evaluation of ERS services and therefore in the ultimate success of ERS as a research and information agency in support of public economic policy.

Timeliness

Although timeliness is an important attribute in staff analyses, information, and research, the nature of its importance is quite different in each of these three functions.

Staff analyses bring information to bear on very specific policy questions as they arise in real time. ERS, like all agencies in support of economic policy, stands ready to provide staff analyses on demand as they are required. These analyses may be requested anywhere from a day to several weeks before the chief economist or the secretary of agriculture testifies before a congressional committee, before the secretary makes a speaking tour, or immediately following a cabinet meeting. In some cases, the agency may literally need to provide information on a moment's notice—say, to answer the secretary's questions about a written brief as the secretary rides to the airport. Deadlines for staff analyses are typically clear, and it is very unusual for ERS or any similar agency to miss a deadline. The quality, relevance, and credibility of the information provided are the dimensions on which these analyses are more likely to vary.

Information and secondary data are provided by ERS on a regular schedule. The development and publication of indicators and the publication of *Situation and Outlook* reports are all subject to known deadlines. Determining the frequency with which this sort of information should be provided, as well as the level of detail, is an important management concern for ERS.

The timeliness of research is an extremely important attribute in any agency in support of public economic policy. Policy decisions are made on a daily basis. The evolution of critical policy questions is never perfectly predictable and often difficult to discern. Regardless of how predictable or unpredictable the policy process is, the specific decisions that actually constitute policy are often made under great time constraints. When those in policy support positions, for example agency heads, are asked to provide information pertinent to these decisions, there is no time to synthesize pertinent research, much less carry out the research itself.

Since research in support of policy questions cannot be conducted on a just-in-time basis, those providing research in support of public economic policy must do their best to anticipate these questions and direct a program of research that is most likely to provide the best support when questions arise. When a research director can anticipate the timing and content of policy questions with accuracy, the course of action is clear. Ideally, the agency can then respond to the question, at the time it is posed, with a dispassionate report meeting high analytical standards prepared before the heat of debate. In the more common case of highly imperfect foresight, those guiding research must seek to position research projects or syntheses of research and data development so that, taken together, they provide an up-to-date knowledge base in the important subject matters on which the agency's policy responsibility and mission depend.

Relevance

The programs of a research and information agency in support of public economic policy must be driven by the needs of its clients. In the case of ERS, the chief client is the secretary of agriculture. An effective research and information program provides the infrastructure that enables staff analyses to meet high standards for analytical expertise and accuracy in directly answering the questions posed. Achieving and maintaining relevance in ERS or any similar agency is the product of a series of accomplishments demanding intellectual and managerial skills of high order.

Since research programs and the development of good secondary data require significant lead times, it is essential for those managing research and information programs to be in constant contact with public and private decision makers. This can be done directly, through personal interaction and reading, and indirectly, by following the critical press. Advisory panels of outside experts and regular contact with foreign counterparts and industry leaders can also play a role in keeping fully abreast of emerging and future policy issues.

Determining the appropriate mix of programs in the light of evolving policy questions, budgetary constraints, and available skills is fundamental to maintaining the relevance of research and information programs. Feedback from staff analysts to managers can be an important ingredient in assessing and maintaining the relevance of programs—in determining, for example, information and research that would regularly improve responses if it were available, and in establishing that information once widely demanded and still produced is no longer used in internal staff work.

Having identified emerging or future policy issues, identifying the research required to inform the likely policy decisions requires keen analytical skills on the part of managers. Framing the appropriate question and recognizing the institutional, political, and other dimensions of the policy context in the economic analysis are critical to relevant research. Not only can mistakes at this stage produce research that is irrelevant, but also the irrelevance of the research may further obscure the real issues when the time for a decision arrives. It may also lead to substantial investments in new research, when in fact answers to the appropriate question may have already existed in the literature and only a synthesis of existing results was called for. With reference to the models for research and information agencies in support of public economic policy discussed above, this task is regarded as being of singular importance by the managers of ABARE and the PAE office of DOD.

Quality

The research and information program of an agency in support of public economic policy can be judged by disciplinary and professional standards. In the case of ERS, these will often be the standards of the discipline of economics for research and that of statistics for providing secondary data, as well as the professional standards of applied policy analysis. For many services, the point of evaluation for the client is most likely to be a comparison of the quality of the ERS product with that of another, actual or hypothetical, supplier of the service, subject to similar constraints on timeliness, budget, and relevance of the response to the question posed. For managers in a research and information agency, there is therefore an implied efficiency standard of quality in all of the agency's work: for a given allocation of budget and a specified time period, and for the given analytical framing of the question, it should not be possible to obtain higher-quality research and information than that currently being delivered.

An ongoing task of any agency is to examine its system for procuring research and information with respect to this efficiency standard of quality. Since disciplinary knowledge and standards constantly advance, current knowledge must be incorporated in the provision of research and information. For intramural work, new knowledge can be infused by hiring new, recently trained staff, by employing visitors with up-to-date skills, and by continuing the education of existing staff. For extramural work, meeting current disciplinary standards should be made an important part of the scoring function for the award of contracts and grants.

The standards of publication in peer-reviewed academic journals are at best indirectly related to the attribute of quality of research in a research and information agency in support of public economic policy. Although regular journal publication of research provided by the agency is a reliable indicator of the quality of the research, it may also reflect an undue emphasis on methodological innovation at the expense of relevance. Failure to publish in such journals may indicate that, in addressing a relevant question in a timely way, methodological innovation was not necessary or was not possible given constraints. Using such standards fails to recognize that the appropriate criterion is one of comparison: given all of the other relevant constraints, is the research and information delivered of the highest attainable quality? Meeting the latter, relevant standard of quality is an important key to successful evaluation of the agency's research program, and to its credibility and reputation in the long run.

Credibility

Establishing and maintaining the credibility of a research and information agency in support of public economic policy is in the long-term interest of the agency and of the department of which it is a part. Credibility within the department is primarily the responsibility of agency management. Credibility beyond the department requires the support of the secretary as well as appropriate management of the agency.

Within the department, senior agency managers, or their designated liaisons, must be able to answer questions posed by decision makers quickly and in plain language. This requires a thorough understanding of the question being posed, the framing of the relevant research questions, and how well agency research and information respond to these questions. These qualities are distinct from those the agency seeks in its senior scientists and research investigators. If the agency regularly meets high standards for timeliness, relevance, and quality, then its findings will have a respected seat at the table when decisions are made. A reputation for such standards makes it more likely that agency findings that contravene the policy position of USDA or the administration will be given a hearing. Conversely, findings that support policy will be regarded seriously, even by those who hold opposing views.

The credibility and reputation of the agency can carry beyond the department. At one extreme, if the agency is known only for providing findings selectively and after the fact in support of positions already established for political reasons, and if the department is known to censor findings that would contravene its position, then the agency's research will carry no weight. Even when dispassionate, anticipatory research in fact supports USDA's position, that fact will be of little advantage for USDA or the administration in dealing with Congress, the public, and other countries.

At the other extreme, if agency research meets high standards for quality and relevance and is made public without regard to established political positions, then it will carry considerable weight outside the department and may be used effectively. As a practical matter, this latter situation can be approached only if research meets the standard for timeliness set forth above—namely, if it takes place in anticipation of a future policy decision, and its credibility is sealed by public delivery of a report prior to the policy debate. To the extent that the agency can demonstrate that it identified and framed the problem for analysis and that political considerations did not enter findings of fact, its credibility will be enhanced. This requires some distancing of those who provide counsel to the secretary from those who carry out the research. In the short run, this insulation of research findings from the political process can be awkward and even on occasion painful, but in the long run—as discussed in Chapters 3 and 4—it is in the best interests of the agency, the department, and the nation.

PRINCIPLES FOR RESEARCH AND INFORMATION ADMINISTRATION

The responsibility of ERS is to provide research and information in support of public economic policy to its clients and in particular to the secretary. In the long run, there are several, mutually competitive, ways that ERS could meet this responsibility. Within the framework for evaluation described in Chapter 5, ERS must examine these possibilities with respect to their prospective impact on the eventual evaluation of its work.

Principle of Competition

The principle that services should be supplied competitively is a fundamental premise of our economic system, including government procurement. It is recognized in the Competition in Contracting Act of 1984, and it has long been the foundation of most publicly sponsored scientific research. Against this principle, there are always powerful forces pushing in the direction of a mandated sole-source supply for services. In the public sector, these forces repeatedly manifest themselves in legislation earmarking the creation of a research institute here or a laboratory there, and in rules and institutions that favor continued relations with established suppliers.

A competitive supplier of credible research services, in particular, must constantly integrate new ideas in order to continue supplying those services. There is pressure for both individuals and organizations to reach beyond their immediate area of expertise to gain a competitive edge by bringing to bear new results from related fields. This process has been a hallmark of the public program of grant research in science and medicine administered in the United States over the past 50 years. Suppliers of research services with permanent, sole-source awards have no such incentives to reach out, and they often become isolated within narrow fields using methods that are increasingly outdated.

As emphasized in Chapter 5, comparison and competition are central to the process of evaluation. Eventually, clients for services will make choices among alternative providers. A successful manager will recognize this fact and look ahead to clients' competitive comparisons in deciding how services will be delivered. To expect any less, in a government agency, would be an abuse of public trust. Managers of research and information agencies in support of public economic policy, in particular, must have the discretion to choose widely from among potential vendors of the services they are responsible for providing to their clients, and they must exercise that discretion. Much of the balance of this report addresses procedures and institutions to facilitate such discretion, while ensuring that the decisions of research and information managers reflect the interests of their clients rather than the interests of vendors.

Potential Impediments to Competitive Procurement

No organization should ever be given a permanent sole-source award for the provision of a service. Decisions to provide sole-source awards must be defended on a recurring basis, beginning from the presumption that services should be procured competitively. Extended conversations with agency heads and former agency heads revealed three reasons why an agency may decide not to undertake competitive procurement.

• Transaction Costs. If the transaction costs of setting up the competition are large relative to the size of the service being procured, then it may be more cost-effective to provide the service on a sole-source basis. The procurement could be from an outside vendor, from another agency, or from within the agency itself. The unit appropriate for procuring services is therefore an important management decision: at one extreme, agencies make small grants and contracts for well under $100,000 for specific deliverables; at the other, agencies make multimillion-dollar periodic commitments to federally funded research and development centers for wide-ranging research over multiyear periods.

• Timeliness. If timeliness is critical, then it may be essential that services be maintained within the agency, to be available instantaneously. This case is most compelling when information is regularly needed on demand at a level of detail that can be provided only at the career staff level, as opposed to economic advisers appointed to serve their client. For example, if the secretary regularly requires information on less than a few hours' notice on such topics as the impact of current drought conditions on farm foreclosures in Texas in the past month, then permanent intramural staff may indeed be the only alternative. However, instant counsel on more general economic matters could be obtained from an under secretary not subject to OPM rules, and research into the impact of weather-related risk on farm finance could be undertaken by a variety of individuals and organizations outside ERS.

• Confidentiality. The need to maintain secrecy may, or may not, be a reason not to close a service to competitive procurement. With respect to the substantive work in ERS, use of privileged information for personal gain is the main concern. Acting on advance knowledge of this new information could yield substantial personal fortunes and is forbidden for the same reason that insider trading is a felony. Before ERS and the National Agricultural Statistics Service release new data and attendant forecasts of crop yields and production, great physical security is provided in the final stage: the so-called lock-up, in which analysts are not permitted to leave the lock-up area in the final hours before the processing and release of the information. Similar concerns are cited by other government agency heads, in matters in which inside knowledge of government action would provide large profit opportunities, or could cause market prices to move against government procurement of goods or services. In most government procurement, however, this concern is well addressed primarily by felony penalties for insider trading and divulging top secret information, which apply to all individuals whether or not they are federal employees.

It is important to emphasize that these three reasons are the only ones agency heads have identified as reasons why a service should *not* be procured competitively. According to the principle of competitive procurement stated here, this means that, when these factors are not germane, then the service *must* be procured competitively and when they are germane, the service *may* be procured competitively.

Application of the principle of competitive procurement, in a given situation, could lead to a number of alternative decisions. The decision might be to solicit bids from private-sector suppliers, or to set up a federally funded research center, or that only ERS employees should provide the service. In the latter two cases, the award would be sole-source, but it would not be permanent. The decision to use a sole-source arrangement must be reconsidered periodically, and every sole-source supplier must be made aware of this fact. The only permanent sole-source award in a democracy is to the electorate for choosing its government.

RECOMMENDATION 6-1. Research and information in support of public economic policy should be procured competitively. All potential suppliers, including ERS, should be on the same competitive footing. If an outside supplier is selected as an awardee, in many cases ERS should have a secondary role as a partner in the provision of the service. No supplier, including ERS, should have a permanent, sole-source award for the provision of any service. Any decision to grant a sole-source award must be defended periodically.

The partnership role of ERS is critical to the effective competitive procurement of research and information in support of public economic policy. It is essential that USDA ensure that the supplier has an understanding of the policy context for the research or information and apprise contractors of developments in policy over the lifetime of the award. Intellectual command of the areas in which research is contracted, as well as an understanding of the universe of potential suppliers and their capabilities, are essential to making appropriate decisions about how research and information will be procured. It is also important that continuity be maintained when there is a change in suppliers, and vital that the research and information procured be readily available to meet demands for staff analysis. In many cases, a partnership role for ERS may be the most effective way of meeting these goals. The relationship between ERS, policy makers in

USDA, and the providers of research and information, is likely to vary with the characteristics of alternative suppliers of research and information services and with the nature of their research.

Competitive procurement can go far to address two of the most important aspects of the widespread perception that the quantity and quality of ERS products are not what they should be, given the number of professionals in ERS and the size of its budget. The first factor, identified in Chapter 4, is that the USDA mandate in economic policy demands a broader range of professional skills than is now present in ERS or is ever likely to be present in the permanent staff of an intramural research agency. Moving away from the permanent sole-source award model means that the best qualified professionals are candidates in providing research and information needed to inform policy decisions. The second factor, also identified in Chapter 4, is that salaries and other dimensions of career opportunities in ERS are less attractive relative to alternatives than they were at one time. USDA and ERS have limited discretion over salaries and some dimensions of career opportunities for their own employees, but through competitive procurement they have access to a much wider universe of professionals, many of whom are not subject to the same limitations on salaries and career opportunities.

Alternative Suppliers of Research and Information Services

The evaluation process set forth in Chapter 5, which forms the basis for administration of research, information, and policy analysis services, entails comparison of suppliers and potential suppliers of services with respect to service attributes valued by clients. At the inception of the BAE in 1922, these services could be supplied only by an intramural research agency. There are now several alternatives.

Private Sector

Chapter 2 noted that, if information is nonrival and nonexcludable, then it is a pure public good. Most of the information produced and used by ERS is nonrival—that is, it is not diminished when it is used by an individual or organization. The information is also excludable—that is, individuals can be charged for their use of it: for example, abstracts of ERS reports are available at no additional cost through the Internet, but copies of full reports require payment. Similarly, the information contained in *Situation and Outlook* reports is nonrival but is excludable and is therefore not a pure public good. What information, if not produced by ERS, would be produced in the private sector? The answer to this question is surely neither none nor all. Changes in technology have substantially lowered the costs of producing secondary information and have greatly increased the options for making it widely available but excludable. The possibilities for delivery to a large group of clients at positive but low cost have developed rapidly. In the light of these developments, the questions of user fees and the likely structure of the market for the kind of information ERS provides, if it were to either charge user fees or withdraw completely, needs to be examined carefully and dispassionately.

Intramural Research Agency

ERS provides research and information almost solely from its staff of about 550 employees, including over 300 professionals.[1] Other research agencies use a combination of intramural and extramural research—this is true even of the National Security Agency, which operates a small extramural research program. This report documents the important global considerations in the choice of vendors, but it does not enter into this choice at the service-by-service level appropriate to actual management. On a service-by-service basis, ERS might benefit from the experience of other agencies in their distribution of work between intramural and extramural vendors.

Federally Funded Research and Development Centers (FFRDCs)

Thirteen federal departments or agencies utilize 37 FFRDCs (NSF, 1998b). In each case, the research sponsor and the FFRDC enter into a relationship that is long term but not permanent. The relationship is reviewed periodically. Substandard performance can in principle lead to the termination of the relationship. FFRDCs develop and maintain models, surveys, and other infrastructure in support of the mission of their client agency. They hire professionals who often spend major portions of their careers in the FFRDC. Many FFRDCs compete directly with the very best academic institutions to retain key senior personnel. They are able to move personnel between assignments and on occasions respond rapidly to short-term requests, in ways that universities and university-based research institutes usually cannot. They are not subject to OPM rules. The relationship between an agency and an FFRDC requires careful management. The agency and the FFRDC must establish a working, long-term relationship. At the same time, this relationship cannot degenerate into a permanent sole-source contract, with personnel moving back and forth between the agency and the FFRDC. The possibility that the FFRDC would lose a contract given substandard performance must be real.

---

[1]A recent departure from this mode of operation took place in March 1998, with the request for proposals "Food Assistance and Nutrition Research Program: Studies of Households Who Leave the Food Stamp Program." This initiative was undertaken as a consequence of the 1998 appropriations bill for USDA, which consolidated all research and evaluation studies of the USDA's food assistance programs under ERS and increased the budget for these studies.

University-Based Research Institutes

University-based research institutes have some of the same characteristics as FFRDCs. They have long-term relationships based on multiyear contracts with the sponsoring agencies, but at regular intervals—at least 5 years, rarely more than 10—there are serious recompetitions that the incumbents lose with some regularity. Their work is circumscribed by the policy responsibilities of the sponsoring agency. Compared with FFRDCs, the emphasis is more on long-term research, less on the development of infrastructure, and university-based research institutes rarely provide direct support for day-to-day staff analyses. They typically draw heavily on the academic expertise of their university base and employ flexible arrangements to involve outstanding senior scholars on a part-time basis. The relationship between an agency and a university-based research institute requires careful management. Although day-to-day involvement in policy is not the reason for using a university-based research institute, its research must ultimately be relevant to the policy responsibilities of the agency. The agency's sponsorship must not simply become scholarship and stipend support for open-ended academic research.

Grant and Contract Research

Grant and contract research can be used to bring concentrations of specific talent to bear on particular research questions. The universe of potential investigators is greater than under any of the other modes of procurement, and specific individuals or groups may be matched well with specific research questions. Success in using grant and contract research requires careful attention to the framing of the question, the solicitation for proposals, the evaluation of proposals, and the maintenance of long-term relations with vendors.

The framing of the question is critical. As discussed previously, identifying the research questions relevant to the anticipated policy question requires skill and careful consideration. Agency research administrators often draw on the expertise of agency clients and potential vendors in framing the question. This can serve to familiarize vendors with the pertinent policy questions and provide clients and administrators with some idea of what research may be able to contribute to policy analysis within time and resource constraints. Occasionally agencies have held short conferences for this purpose. As administrators develop ongoing relations with the community of potential vendors, they improve their own management decisions by drawing on talent beyond their agency.

The proposal solicitation must make clear the terms of evaluation to prospective suppliers. For example, if the research being procured is in an area in which there has been considerable academic work but a dearth of findings relevant for a particular kind of policy decision, then this must be made clear in the solicitation so as to avoid the impression that a reworking of the academic literature will suffice. The solicitation should clarify the relative weight attached not only to the proposal itself, but also to the characteristics of the vendor. In particular, substantial weight should be given to the past performance of the proposer, if the proposer has previously been a vendor for the agency. This provides a proper set of incentives for the supplier who is selected to carry out the research, for once this is done the supplier is in a monopoly position with respect to the agency for that project. In general, the evaluation criteria for a research and information agency in support of public economic policy will be quite different from the criteria used by agencies that fund external basic research, including the National Institutes of Health, the Agricultural Research Service, and the National Science Foundation. These agencies tend to emphasize quality with loose restrictions on the area of research, whereas for an agency like ERS the relevance to a specific policy context and timeliness of the work is often critical. Wholesale adoption of basic research procurement models is bound to be inappropriate.

In most cases, the evaluation of proposals involves considerably more than just selection by agency administrators. The same group involved in framing the question will typically have important contributions to the evaluation of proposals—including knowledge of agency clients, qualified experts, and potential vendors free of conflict of interest. The best proposals often raise issues that were not foreseen at the time the solicitation was drawn up: new ways of framing the question may be suggested, unforeseen difficulties with methodology or data may be uncovered, and new and untried techniques may be proposed. In some cases, the agency may obtain substantially improved research by reframing the proposal on the basis of this new information, and it may wish to enter into agreements with more than one vendor.

It is important to develop long-term relationships with good suppliers. On one hand, good suppliers should be able to rationalize a long-term investment in skills specific to agency needs, knowing that there will not be capricious decisions about programs and vendors. This is especially important when the work demands that those doing the research make specific investments in a project that cannot be transferred to other projects. On the other hand, all suppliers should understand that all work is subject to recompetition, and that poor performance will result in the loss of the contract.

RECOMMENDATION 6-2. Choices among alternative vendors of research and information in support of public economic policy should be based on prospects for favorable evaluation of the services provided, as well as on the costs of the services. The critical attributes established in program evaluation provide the framework for choice among vendors. No single model of choice among vendors is appropriate for all programs. In particular, the methods used by the National Science Foundation, the National Institutes of

Health, the Agricultural Research Service, and the National Research Initiative will not be suited to many ERS programs and should not be presumed to be appropriate to any.

ORGANIZING RESEARCH, INFORMATION, AND POLICY ANALYSIS

Many of the considerations taken up in this chapter apply to most research and information agencies in support of public economic policy. There are specific features of ERS and its history that bear on the application of these principles in organizing its research, information, and policy analysis services.

ERS in the Wider Research and Information Community

For most of its history ERS has been closely tied in many ways to the land grant institutions and the departments of agricultural economics in those universities. From the late 1920s to 1983, when some ERS professional staff were posted outside Washington, most postings were to land grant institutions. The research and information mandate of ERS, as reflected in Box 5.1, still includes quite a few topics found in the portfolio of the traditional agricultural research establishment, but many other topics—financial institutions, nutrition, climate change, auction mechanism design, price indices, food assistance, etc.—are widely studied by economists and other social scientists in many universities, research institutes, and the private sector. The principles of evaluation set forth in Chapter 5 imply that ERS should develop relations with a wider community and in its research program it should encourage new combinations of ideas from all pertinent disciplines.

RECOMMENDATION 6-3. In moving to a system of competitive procurement from a system of permanent sole-source retainers, the long-term commitment of ERS to competition should be conveyed to all potential suppliers of services. If ERS seeks to develop long-term relations with the entire community of potential suppliers, then this fact must be conveyed credibly.

USDA and the land grant institutions have strong ties with Congress dating to the time when a majority of Americans lived in rural, rather than urban, areas. Many of these ties remain today. Congress makes direct appropriations to USDA for state research, extension, and forestry, which USDA in turn distributes to the states according to congressionally defined formulas. The extension system, in turn, extends literally to every county in the country. The early rationale for this system can be traced largely to the reluctance and inability of states to invest in research, the location-specific nature of agriculture, the diversity of agricultural markets and products, and the historically atomistic nature of production in agriculture. It has played a major role in the abundance and low cost of agricultural products in the United States. But these same factors worked against competition between suppliers of research service and their application. The necessity for genetic adaptation to specific ecosystems dictated that hybrid corn appropriate for Missouri would not compete with that developed in Michigan, and pest management in California could not be taken wholesale to Wisconsin. The politics of agriculture reinforced unconditional commitments of research funding to states and institutions.

In the period since World War I, state-level investments in agricultural research have grown to exceed that of the federal government, creating a very different incentive for federal appropriations to states (Alston and Pardey, 1996). A significant portion of the benefits from any one state's research investment now spill over to other states, reducing the incentive for states to invest to an optimum level nationally (Alston and Pardey, 1996). Exceptionally high levels of return to agricultural investment persist, supporting the conclusion that the nation has been and continues to under invest in such research (Alston and Pardey, 1996; Huffman and Evenson, 1993). Without adequate compensation to the states for their loss of benefits, this will continue.

This decentralized model is less applicable to the substantive economic research mandate of ERS in the 1990s and the economic policy questions that will confront the secretary of agriculture or the functional equivalent in the foreseeable future. It is essential that ERS reach out to a wider community of research and information providers. In moving to a model in which the presumption is in favor of competitive procurement, it is essential that the competitive mode be taken seriously by the secretary, the executive branch, and the Congress. A permanent sole-source contract has no virtue simply because it is granted externally.

Although long-term, consistent support for the principle of competition and the importance of the attribute of quality must start in the secretary's office, it must also be sustained by similar understanding of the commitment in the executive branch and the Congress. This requires some agreement among these parties on a common set of expectations and rules to govern the role and performance of ERS. Currently this understanding is at best incomplete, with ERS sometimes caught between inconsistent expectations.

RECOMMENDATION 6-4. Support for the principle of competition and the necessary attributes of quality, credibility, timeliness, and relevance by the secretary's office, the executive branch, and the Congress is essential in moving to a system of competitive procurement. It is important to fund projects and people rather than institutions, and to do so subject to periodic evaluation by qualified reviewers.

Mutual understanding by political leaders in the executive branch and the congressional leadership of both parties of what is entailed by quality, credibility, timeliness, and relevance is essential to the effectiveness of ERS or any agency conducting research in support of policy analysis. If the expectations of the executive and congressional branches with regard to these points conflict, then the work of the agency will be less than effective and the budget of the agency is very likely to suffer. Lack of a clear set of common expectations has plagued the performance of ERS for some time.

Staff Analysis

Staff analysis is the link between the ERS research and information programs and the policy decisions that these programs support. Close contact between staff analysis leaders and policy makers is required to ensure that the entire ERS program remains relevant to the substantive economic policy mandate of USDA. Information that must be provided on a very short-term basis—often a few days or less—requires that those providing the information be immediately available.

Staff analysts must also be closely involved in guiding the ERS research and information program, including the assessment of future policy questions, the framing of questions for investigation, and the organization and supervision of research, because staff analysts are the first line of contact with policy decisions. Thus, leadership in staff analysis requires a sophisticated combination of analytical and management skills.

The important attributes of credibility and relevance in staff analysis and the need for staff analysis leadership to participate in overseeing research and information programs indicate that this function must be provided by a permanent, skilled group within ERS.

RECOMMENDATION 6-5. To provide immediate support for its economic policy decisions, USDA should maintain a permanent core of staff analysts. The size and composition of this group should reflect the level of detail and timeliness required in support of the economic policy mandate of USDA, and it should be reviewed from time to time as the mandate evolves. The leadership of this group must provide a combination of management and analytical skills essential to the administration of the research and information programs of ERS. ERS should regularly invigorate this group by means of visiting scholars, sabbaticals, internships, or similar programs, to maintain the contact of staff analysts with the wider research community.

Information

Historically there has been consistent demand for the information services of ERS (and the BAE preceding it). These services include *Situation and Outlook* reports and other information combining some analysis with data. (A more detailed description is provided at the start of Chapter 5.) These services began with a focus on commodity markets but are being slowly extended to support public and private decision making in newer areas of USDA responsibility, including the environment and natural resources. Much of the strongest support for these services has come from smaller users, many of whom have limited analytic capacity to interpret data. There is a demonstrated private-sector demand for these information services, which will remain even if traditional price support programs vanish. There will also be other public policy uses for this information.

Essentially all of this information is both nonrival and excludable. It is being produced by ERS, but not because it is a pure public good. Clients for this information include private-sector decision makers and other branches of government, as well as the secretary. Extensive price support programs for a large number of agricultural commodities require detailed projections of supply, demand, and prices in order to estimate the budgetary and farm income implications of price supports. Under the 1996 legislation these programs are being changed, with the possibility of further reductions or even elimination in 2002, the next farm legislation renewal date. If the programs are eliminated, then much of the secondary information produced by ERS will not be as vital to USDA as it once was for budget forecasting.

However, there could be other reasons for maintaining public provision of this information. The regulation of agricultural production for environmental and other purposes is increasing and could require this type of secondary information and forecasts to analyze the costs and benefits of regulation on agricultural markets and income before regulations are implemented. Also, if there are market inefficiencies caused by asymmetric information (as discussed in Chapter 2), then public provision of market data and forecasts may improve market efficiency.

Different types of secondary data preparation and analysis may also be needed as public priorities change. For instance, with food and nutrition program spending far exceeding government spending on farm price supports, more secondary data and analysis on food consumption patterns and expenditures may prove valuable in examining future changes in food stamp and child nutrition programs.

The attributes of these secondary data and information important for policy analysis within USDA and other government agencies and to other clients would then drive the analysis of the question: Should this information be produced internally at ERS, should it be produced by other vendors under contract to ERS (as indicated in Recommendation 6-1), or should it be left to the private sector? The frequency and detail with which this information should be produced is also subject to prospective evaluation within the framework set forth in Chapter 5.

RECOMMENDATION 6-6. The secondary data preparation and analysis programs of ERS should be evaluated within the framework outlined by the panel, including consultations with clients. On the basis of this evalua-

tion, a long-term plan should be drawn up, including new and discontinued services. The plan should indicate which of the services provided will be produced in ERS, which will be procured from other vendors, and which will be left to the private sector. The plan should include the anticipated impact on clients and the projected impact on the USDA budget.

Research

In the evaluation framework, the attributes of services are the dimensions along which comparisons are made. For ERS services generally, and for its intermediate and long-term research in particular, these attributes are likely to include timeliness, relevance, quality, and credibility. The previous discussion has indicated how these attributes are interwoven in the case of research. In particular, research that is undertaken well before a policy debate reaches a crescendo, and thus is available before policy positions are formed, is more credible than research produced just in time by one of the parties to the debate. The same is true of research that is conducted by the best-qualified investigators, directly addresses the policy issues at hand, and is not subject to clearance by either political or private interests.

An arm's length relationship between the sponsor of intermediate and long-term research and the investigation process itself has important ultimate advantages for the sponsor and other policy makers, as well as for the taxpayers who make the work possible. Clear and public framing of the questions for investigation, including perhaps key assumptions to be made, followed by unimpeded skilled scientific investigation, provides the most propitious environment for research in support of informed public economic policy. If there is to be rational dialogue on questions of economic policy, especially in a charged, partisan political environment, some common ground rules for fact finding are in order in both USDA and the Congress. ERS, and other research and information agencies in support of public economic policy, can do much to bring this about. It is in their long-term interests to do so.

RECOMMENDATION 6-7. USDA should support the integrity of its intermediate and long-term research programs in support of economic policy, while retaining the prerogative to disagree with research findings. These programs should be conducted with the clear objective that peer-reviewed research findings may be published by the investigators independently and without prior approval by USDA, and with the clear understanding that USDA does not necessarily endorse the findings of any research program.

The attribute of quality, discussed at length above, requires that ERS consider the widest feasible group of vendors to provide intermediate and long-term research and to use imagination in the vehicles for this research. For example, in anticipation of an emerging policy issue, ERS can convene one- or two-day meetings in which the relevant issues are conveyed to groups of highly skilled investigators, and ERS managers and staff analysts can then further pinpoint the analytical issues for intermediate or long-term research. These same groups, or individuals within these groups, can also be engaged by ERS as peer reviewers of the scientific content of intermediate and long-term research. Such review would be an important component of quality control, and would be consistent with the support for the integrity of these programs addressed in Recommendation 6-7. ERS, or any other research and information agency in support of public economic policy, can bring more skilled and varied talent to bear at critical junctures in its work, than it could ever contemplate retaining as permanent staff. The attribute of quality points strongly in the direction of using external vendors for most long-term research.

The attribute of credibility reinforces this conclusion. The advantages of arm's length relationships for reaching critical or sensitive conclusions are well understood. Blue-ribbon panels, independent fact-finding investigators, and studies by the National Research Council all underscore this understanding. By contrast, intramural research conducted by permanent staff in a government agency subject to clearance, and studies by private firms carried on internally or by consultants for hire face overwhelming odds against their credibility in any politically charged policy debate. Research in support of public policy depends for its credibility, in great part, on its conduct by those with a greater interest in the quality of the work than in the substance of the conclusion. If ERS can credibly assure professionals—whether permanent employees or external vendors—that their work will not be subject to political interference, then it can more readily attract top talent, thereby further increasing the quality of its product.

RECOMMENDATION 6-8. Vendors for intermediate and long-term research programs in support of economic policy should be sought from the widest possible universe of qualified investigators and organizations. Intermediate and long-term research conducted by all vendors, including ERS staff, must be subject to the understanding that their peer-reviewed research findings may be published without prior approval by USDA.

As previously discussed, this organization of research requires ERS managers and staff analysts to ensure that their investigators understand the relevant policy questions motivating the research. This includes knowledge of institutions, the policy context, and data, as well as methodology. These attributes of good research can be made clear in solicitations and evaluations. (These requirements must be met whether research is conducted internally or externally.) The advantages with respect to quality and credibility of research that is administered internally but conducted externally outweigh these and other transaction costs in most cases.

CONCLUSION

The economic policy mandate of the USDA has grown, in the twentieth century, from issues of specific economic concern to American farmers, to complex and critical economic questions involving social welfare, food security and safety, environmental change, international relations, and other issues affecting everyone in the world. The Department of Agriculture and the Economic Research Service have an opportunity to reach out to a correspondingly wide research and information community to bring to bear the best minds in pursuit of dispassionate solutions to these problems. This requires both a reconsideration of the deployment of information and research resources in support of public economic policy and a reaffirmation of this commitment of resources as a matter of policy. In this context, ERS can make key contributions in addressing these issues in the next century.

Implementing the principle of competition is an important initiative USDA and ERS can undertake in meeting the challenges faced now and in the next decade. The research and information required to support the rapidly widening and challenging policy mandate of USDA cannot be produced to high standards by any single permanent staff of professionals, including ERS. Flexibility in selecting providers of research and information products is essential. When ERS staff are selected, the principle of competition will ensure both the actual and perceived quality and professionalism of the work. Consistent with the principle of competition, a consistent policy of insulation of economic research and fact finding from political intervention will make ERS more attractive to motivated and capable professionals, and will further improve perceptions of the quality of ERS products.

7

Organization and Placement

The appropriate administration of services, taken up in the previous chapter, will be effective in delivering research and information to policy makers, in particular the secretary, only if it is embedded in an organization that supports it. Both the internal organization of the Economic Research Service (ERS) and the organization of the U.S. Department of Agriculture (USDA) with respect to ERS need to be considered. This chapter draws from the administrative model developed in Chapter 6 and the lessons of experience recounted in Chapters 3 and 4, to propose effective organization for research and information in support of public economic policy in USDA.

MANAGING THE FORM, SCOPE, AND SIZE OF ERS

ERS is an organization that draws together data, other information, and economic analysis to produce concise factual information that is immediately useful in informing policy makers, chiefly the secretary, of policy alternatives and their consequences. To do this effectively, it must embody an understanding of policy issues, while reaching out to a wide universe of potential sources for data, other information, and economic analysis. Maintaining both the internal resources for timely delivery of relevant concise information and the external scope of potential vendors needed to ensure the quality and credibility of information and analysis are the functions that drive the form, scope, and size of ERS.

Ingredients for Effective Policy Analysis

An effective research and information agency in support of public economic policy must acquire primary data, produce secondary data and other information in a form that is useful for both quick analyses and longer-term research, and provide research that improves current and future decision making. The acquisition and organization of information, on one hand, and the conduct of long-term research, on the other, are intimately bound together by policy problems. Effective management maintains the relevance of both information organization and research. Effective research requires the right information, and information should be produced only if it is going to be used.

Producing primary data is a very small part of ERS activities. The only significant ERS responsibility for primary data is the Agricultural Resource Management Study survey, which is actually fielded by the National Agricultural Statistics Service (NASS). Most primary data used by ERS come from NASS, USDA program agencies, the Bureau of the Census, and the Bureau of Labor Statistics.

Secondary data preparation and analysis, including *Situation and Outlook* reports and indicators, are a very significant part of ERS activities, accounting for about 40 percent of the time of ERS professional staff. Executive and congressional decision makers and a large set of private-sector decision makers use ERS secondary data and analysis. Responses to routine queries typically amount to staff's obtaining the relevant tables and summaries and explaining how they answer the question. This sort of activity accounts for up to 20 percent of the time of ERS professional staff. The staff engaged in producing secondary data frequently have very detailed knowledge of the institutions and markets to which the data pertain, although, given the reduction in the personnel and resources of ERS, the level of detail and redundancy available to ensure continuity in quality are not what they once were.

Producing long-term research is also a very significant part of ERS activities. In some cases, ERS research entails economic analysis of important policy questions being addressed or likely to be addressed by USDA. The economic analysis involves isolating the economic essentials of the problem at hand, making appropriate and supportable assumptions about the economic behavior of the parties involved, reaching conclusions about the effects of alternative policies, and expressing the appropriate qualifications and uncertainty about the conclusions. In other cases, ERS research reports are descriptive presentations of secondary data. These reports are valued by a broad range of users as potentially useful inputs to policy analysis, and they can do much to facilitate effective economic analysis, but in themselves they do not provide analyses of the likely effects of alternative policies.

Form of Policy Analysis

The need to produce secondary data and research in support of effective policy poses significant challenges to the organization of ERS. On one hand, knowledge of institutions, understanding the policy process, good primary and secondary data, and high-quality economic analysis are all essential to effective policy. Those responsible for policy in a particular area must therefore have access to data, institutional detail, and economic analysis. On the other hand, the professional skills required for good economic analysis are not the same as those necessary for construction of secondary data and a complete grasp of institutional detail. An environment in which data set construction is the main activity is unlikely to be a fruitful milieu for analytical economics, and vice versa. The standards of evaluation for these different kinds of work are quite different, as well: in particular, it makes little sense to evaluate those responsible for secondary data construction by the academic standards that are rightly applied to analytical economic research.

Most ERS projects require interaction between secondary data construction, institutional knowledge, and economic analysis. This requirement is appropriately reflected in the current structure of ERS. The characteristics and career paths of staff professionals involved in these three activities are distinctly different, however. It is quite rare for any one individual to be competitive in both secondary data development and economic analysis; it is generally unproductive to obtain more economic analysis by encouraging career changes for those who have specialized in secondary data development or institutional and market information; and the characteristics and career paths of staff professionals involved in these three activities are distinctly different. Although not equally expert in all areas, a significant number of professionals participate in more than one category of ERS service.

RECOMMENDATION 7-1. ERS management should consider flexible professional staffing arrangements, including the use of visiting scholars and postdoctoral appointments, to obtain the best internal staff.

Scope and Structure

The substantive economic policy mandate of USDA, combined with the process of evaluation applied to the administration of ERS services, determines the scope and structure of ERS. A thorough evaluation of ERS, along the lines anticipated in this report, could greatly change the services supplied by ERS over the long run, and therefore the internal structure of the organization. We present two, possibly extreme, examples for purposes of illustration. Suppose that the program of price supports and similar subsidies currently being reduced is reconstituted in a very different way that requires detailed central management across many commodities. Then at current staff levels, ERS would likely be hard-pressed to provide the information needed to administer this program, even if it abandoned most of its other activities, at current staff levels. It is likely that most of this work would be undertaken internally, and there would be few resources available, at current funding levels, for either intramural or extramural research.

At the other extreme, suppose that commodity-specific economic policy is abandoned entirely, and that USDA's economic policy mandate for social welfare, environmental, and consumer issues continues to expand in scope. The outcome of this change in mandate, and an evaluation indicating that there is little reason for public-sector secondary data preparation and analysis, might produce a structure in which ERS services would best be produced by a core staff analysis group and substantial commitments to support of external long-term research. In this case, too, evaluation might well indicate that research resources should be concentrated with greater intensity on a smaller number of research questions likely to be key in informing future policy decisions.

The future course of the USDA mandate surely lies somewhere between these two extremes. These illustrations are raised here to underscore the fact that the organization of economic analysis in support of policy within USDA derives from the USDA policy mandate. As this mandate evolves, the organization must be reexamined.

RECOMMENDATION 7-2. The appropriate scope for ERS activities should be determined by the economic issues within the policy mandate of USDA, by the system of program evaluations described in this report, and by implementation of the principle of competitive supply of research and information in support of public economic policy.

Taking into account the greatly increased complexity of the policy issues in USDA, the very substantial reductions in ERS staff, and the growth in alternative sources for policy analysis, the outcome might well be that ERS should concentrate on a smaller range of issues but with no decrease in real resources. Proportionate changes in scope and resources, regardless of direction, are unlikely to be productive.

There is a gross inconsistency between the declining real resource base of ERS, the growing diversity and scope of USDA policy information needs, and the conflicting priorities and expectations of ERS clientele, including the secretary and Congress. This problem, which ERS cannot itself solve, must be faced by these parties jointly.

ERS WITHIN USDA

Throughout the history of ERS and its predecessor the Bureau of Agricultural Economics (BAE), the position of the agency within USDA has changed several times. The current position of ERS within USDA is not conducive to its mission to provide research and information support for the economic policy mandate of USDA, relative to either past or potential arrangements. To provide improved policy support and to implement the recommendations for changes in the administration of ERS services made in this report, reorganization of ERS and the Office of the Chief Economist in USDA is necessary.

Current Lines of Responsibility

The current organization of USDA, with respect to economic policy and the research and information support of economic policy, is presented in Figure 4.1. The Office of the Chief Economist is situated in the Office of the Secretary of Agriculture. The chief economist has direct contact with the secretary in policy meetings and has a small staff of about eight professionals. Several other specialized units report to the chief economist (see Figure 4.2). In contrast, the administrator of ERS reports to the under secretary for research, education, and economics, along with the administrators of the Agricultural Research Service, the National Agricultural Statistical Service, and the Cooperative State Research, Education, and Extension Service. The under secretary has no responsibility for economic policy, and the disciplinary backgrounds of the individual in this position have been biology, food and nutrition, or education. Many requests to ERS for staff analysis come from the chief economist.

These lines of authority do not well serve research and information in support of economic policy within USDA. The administrator of ERS, with responsibilities for administering over 300 professional employees, is several steps removed from the policy process to which the work of ERS must be relevant. The chief economist, charged with representing economic information in the decision-making process, has no direct line of authority to the greatest concentration of talent in USDA for marshaling this information. There is bound to be competition between the administrator of ERS and the chief economist in providing both the secretary and other decision makers—including pertinent congressional committees—with economic analysis and prospectively assessing the impact of proposed changes in policy.

These lines of authority would not serve well research and information in support of economic policy under the model of competitive procurement of services by ERS advanced in this report, either. In the current organization, there is no position suited to deciding whether particular information and research services in support of economic policy should be procured from outside vendors, or, in the event that both ERS and outside vendors might supply services, whether or not ERS should be chosen. The administrator of ERS, as a potential bidder in most cases, cannot have the final authority for developing solicitations and choosing from among vendors. The Office of the Under Secretary for research, education, and economics, is not designed to do so. For the chief economist to do so would involve interagency fund transfers each time a decision is made. Reorganization of the economic policy support function within USDA should therefore be considered simultaneously with the question of how these research and information services are procured.

Organization of Economic Policy Support in USDA

The principles for procuring information and research, the lessons learned from the history of the BAE and ERS, and the experience of other cabinet-level agencies suggest a reorganization that copes with all of these problems. First, both economic policy decision making and research and information in support of economic policy should be brought into a single line of authority. This was the case for most of the history of the BAE and ERS, and it is true in many cabinet-level agencies today, including the Assistant Secretary for Policy and Evaluation in the Department of Health and Human Services, the Assistant Secretary for Policy in the Department of Labor, and the Office of Program Analysis and Evaluation in the Department of Defense. Second, consistent with the lessons learned from the history of the BAE and ERS and with the model for procurement of information and research services developed in this report, the functional separation between policy decisions, on one hand, and credible research of high quality in support of these decisions, on the other, should be clear and transparent. These considerations lead to the following recommendations, presented in order of increasing specificity.

RECOMMENDATION 7-3. ERS should never be involved in recommending or deciding on specific policy actions, which are the prerogative of the secretary.

RECOMMENDATION 7-4. A small, highly capable policy analysis and advisory group should be led by an appointee, such as a chief economist or assistant secretary for economics, who manages day-to-day economic policy staff support for the Office of the Secretary. Such a unit would be appointed to serve the secretary and would provide any advice on political and policy action, keeping prescriptive advice and highly political matters from being directed to ERS.

RECOMMENDATION 7-5. The administrators of the Economic Research Service and the National Agricultural Statistical Service should report to the chief economist or the assistant secretary for economics.

Process of Economic Policy Support in USDA

The Office of the Chief Economist or the Assistant Secretary for Economics, in this recommended organization, requires individuals with a thorough understanding of current and emerging policy issues and strong abilities in framing research questions. The set of skills required for these staff people is quite similar to those needed on the staff of the president's Council of Economic Advisers and in the economic policy support agencies for cabinet departments previously mentioned. Anticipating emerging policy questions and ensuring that outside research is brought to bear on these questions are critical. Staff must have a keen sense of the policy environment and an ability to identify the best researchers for each issue. The Office of the Chief Economist or the Assistant Secretary for Economics must be able to pose well-framed research requests that address their policy needs, while balancing timeliness, qualifications to do the work, and a sense of what is possible.

The professional staff in the Office of the Chief Economist or the Assistant Secretary for Economics, and not ERS, would be responsible for bringing the research and information services of ERS to bear in policy councils. They must therefore have a thorough command of the economics of policy questions, whether provided internally by ERS, through sponsored extramural research, or through syntheses of existing research and information. The same staff of the Office of the Chief Economist or the Assistant Secretary for Economics would be responsible for evaluating the program of research and information conducted externally and exercising oversight to ensure that research and information programs are oriented to clients, and not co-opted by vendors.

The administrator of ERS, in this proposed organization, would be responsible for the administration of internal research and information projects and would have a direct interest in maintaining programs that are competitive with alternatives in the public, private, and academic sectors. The administrator of ERS should be a professional, career economist, not subject to political appointment. He or she would be available to explain the research and information findings of ERS, as would external contractors, but should never be called on to represent the policy position of the secretary, the assistant secretary for economics, or the chief economist.

Achieving an organizational structure that insulates the production of economic information and research from political considerations, while ensuring that this work will be relevant to the economic policy mandate of USDA, is a difficult but essential task. To the extent that this is done well, ERS will be more attractive to motivated and capable professionals as a place to work. This is especially important as ERS reaches out to a wider group of professionals than it has in the past. To the extent that this is done well, the entire culture of ERS will change. Expectations will be raised. It will be more difficult for new administrators and division directors to compromise standards for timeliness, quality, relevance, and credibility or to intervene in research programs for political purposes.

The economic policy mandate of USDA continues to grow in scope and complexity. The secretary's is a political position, and he or she cannot be expected to act otherwise. At the same time, precisely because of the analytical complexity of the issues the secretary confronts, it is essential to have research and information support that meets the highest standards for quality, timeliness, and relevance. In a politically charged environment, clear steps must be taken to ensure that considerations of fact and analysis in policy decisions are credibly conveyed as such. This report has outlined steps that can be taken in pursuit of these goals. Their attainment is essential to maintaining the world leadership of the more broadly defined U.S. food and agricultural sector and to addressing key economic problems facing the United States and the world.

CONCLUSION

This report began by considering the nature of public economic policy and by indicating the enormous benefits that derive from *informed* public economic policy. It has presented several alternative arrangements for the production of information, research, and analysis to inform public economic policy. Both history and analysis indicate that, in the case of ERS, some of these arrangements work well and others do not, and this report provides some reasons why this is so. The report's recommendations, based on this history and analysis, provide a process for finding effective arrangements.

Adoption of the recommendations in this report will be effective only if there is agreement among senior policy makers on the principal points underlying them. These points include the nature of public economic policy and the desirability of informed rather than uninformed public economic policy. In the production of information, research, and analysis to inform public economic policy, they include the principle of competition and the desirable attributes of quality, relevance, timeliness, and credibility.

The operation, even the concept, of an agency that informs policy decisions with credible and relevant information but that is not connected to or informed by decision makers is vulnerable, indeed, it is fragile, as amply demonstrated by the history of ERS. Yet the same history indicates that this role is essential to success in informing policy decisions. The concept of such an agency is too fragile to sustain disparate expectations by the executive and legislative branches. It requires cooperation and agreement between the secretary and the relevant congressional leadership on a common set of expectations and rules for shared access to ERS services and its role and expected behavior in dealing with both branches of government. Only in such an environment will informed public economic policy survive.

References

Abel, M.E.
   1991   The role of ERS in Situation and Outlook work. Economics and Public Service: Proceedings of the 30th Anniversary ERS Conference. ERS, USDA (AGES 9138) (July):30-31.
Aigner, D.J.
   1981   The Residential Electricity Time-of-Use Experiments: What Have We Learned? Chapter 1 in J.A. Hausman and D.A. Wise (eds.), Social Experimentation. Chicago: University of Chicago Press.
Alston, J.M., and P.G. Pardey
   1996   Making Science Pay: The Economics of Agricultural R&D Policy. Washington, D.C.: AEI Press.
American Farm Economics Association
   1930   Candidates for the doctor's degree in agricultural economics in American universities and colleges, 1927-30. Journal of Farm Economics 12(2):518-522.
Baker, G.L., and W.D. Rasmussen
   1975   Economic research in the Department of Agriculture: A historical perspective. Agricultural Economics Research 27(3-4):53-72.
Baker, G.L., W.D. Rasmussen, V. Wiser, and J. Porter
   1963   Century of Service: The First 100 years of the United States Department of Agriculture. Washington, D.C.: U.S. Department of Agriculture.
Black, J.D.
   1947   The Bureau of Agricultural Economics—The years in between. Journal of Farm Economics 29(4, Part II):1027-1042.
   1953   Introduction to Economics for Agriculture. New York: Macmillan Co.
Bonnen, J.T.
   1980   Observation on the changing nature of national agricultural policy decision processes, 1946-1976. Pp. 309-329 in Farmers, Bureaucrats, and Middlemen: Historical Perspectives on American Agriculture, Trudy H. Peterson, ed. Washington, D.C.: Howard University Press.
   1986   A century of science in agriculture: Lessons for science policy. American Journal of Agricultural Economics 68:1065-1080.

Bonnen, J.T., W.P. Browne, and D.B. Schweikhardt
  1996   Further observations on the changing nature of national agricultural policy decision pro-
          cesses, 1946-1995. Agricultural History 70(2):130-152.
Bonnen, J.T., D.Hedley, and D.B. Schweikhardt
  1997   Agriculture and the changing nation state: Implications for policy and political economy.
          American Journal of Agricultural Economics 79(5):1419-1428.
Bowers, D.E.
  1990   The Economic Research Service, 1961-1977. Agricultural History 64(2):231-243.
Browne, W.P.
  1995   Cultivating Congress: Constituents, Issues, and Interests in Agricultural Policymaking.
          Lawrence: University Press of Kansas.
Carson, R.
  1962   Silent Spring. Boston: Houghton Mifflin Publishing Company.
Cochrane, W.W.
  1961   The role of economics and statistics in the USDA. Agricultural Economics Research
          13(3):69-74.
  1965   Observations of an ex economic advisor. Journal of Farm Economics 47(2):447-461.
  1983   The Economic Research Service: 22 years later. Agricultural Economic Research 35(2):29-
          38.
  1993   The Development of American Agriculture: A Historical Analysis, 2nd Edition. Minne-
          apolis: University of Minnesota Press.
Cohen, R.
  1998   Redrawing the free market: Amid a global financial crisis, call for regulation spread. New
          York Times A-1:19.
Cotner, M.L., and W.H. Heneberry
  1991   ERS resource economics work—The first 30 years. Economics and Public Service: Pro-
          ceedings of the 30th Anniversary ERS Conference. ERS, USDA (AGES 9138) (July):120-
          140.
Council of Economic Advisers
  1997   Economic Report of the President. Washington, D.C.: U.S. Government Printing Office.
  1998   Economic Report of the President. Washington, D.C.: U.S. Government Printing Office.
Council on Environmental Quality
  1979-1984 Environmental Quality. Washington, D.C.: U.S. Government Printing Office.
Daft, L.M.
  1991   ERS and rural development: A historical perspective. Economics and Public Service Pro-
          ceedings of the 30th Anniversary ERS Conference. ERS, USDA (AGES 9138) (July):146-
          152.
Davis, C.C.
  1940   The development of agricultural policy since the end of the world war. Pp. 297-326 in
          Farmers in a Changing World: The Yearbook of Agriculture, 1940. Washington, D.C.:
          U.S. Department of Agriculture.
Doering, O.C.
  1991   The Economic Research Service, the public, and the profession: A review and assessment.
          Economics and Public Service: Proceedings of the 30th Anniversary ERS Conference.
          ERS, USDA (AGES 9138) (July):18-22.
Duncan, J.W., and W.C. Shelton
  1978   Revolution in United States Government Statistics, 1926-1976. Office of Federal Statisti-
          cal Policy and Standards. Washington, D.C.: U.S. Department of Commerce.
Economic Research Service
  1977   A Chronology of American Agriculture, 1776-1976. Washington, D.C.: U.S. Department
          of Agriculture.

1986   Chapter 1 in Embargoes, Surplus Disposal and US Agriculture. (AER 564). Washington, D.C.: U.S. Department of Agriculture.

1989   How Level is the Playing Field? An Economic Analysis of Agricultural Policy Reforms in Industrial Market Economics (FAER–239). Washington, D.C.: U.S. Department of Agriculture.

1997   Economic Research Service Economist Position Classification System. Washington, D.C.: U.S. Department of Agriculture.

1998a   Agricultural Outlook (AGO–250). April:62. Washington D.C.: U.S. Department of Agriculture.

1998b   Summary Statistics for Staff Analysis (Internal ERS document).

Edwards, E.E.

1940   American agriculture: The first 300 years. Pp. 256-266 in Farmers in a Changing World: The Yearbook of Agriculture, 1940. Washington, D.C.: U.S. Department of Agriculture.

Englebrecht-Wiggans, R., M. Shubik, and R.M. Stark

1983   Auctions, Bidding and Contracting. New York: New York University Press.

Enriquez, J.

1998   Genomics and the world's economy. Science 28(537):925-926.

Finegold, K., and T. Skocpol

1995   State and Party in America's New Deal. Madison: University of Wisconsin Press.

Freeman, O.L.

1991   Economic Research Service: Guide to the future. Economics and Public Service: Proceedings of the 30th Anniversary ERS Conference. ERS, USDA (AGES 9138) (July):96-105.

Gaus, J.M., and L.O. Wolcott

1940   Public Administration and the United States Department of Agriculture. Chicago: Public Administration Service.

Gilbert, J., and E. Baker

1997   Wisconsin economists and New Deal agricultural policy: The legacy of progressive professors. Wisconsin Magazine of History 80(4):280-312.

Guéhenno, J.M.

1995   The End of the Nation State. Minneapolis: University of Minnesota Press.

Hardin, C.M.

1946   The Bureau of Agricultural Economics under fire: A study in valuation conflicts. Journal of Farm Economics 28(3).

1952   The Politics of Agriculture: Soil Conservation and the Struggle for Power in Rural America. Glencoe, IL: The Free Press.

Hathaway, D.E.

1963   Government and Agriculture. New York: Macmillan Co.

1974   Food prices and inflation. Brookings Papers on Economic Activity 1:63-116.

Heinz, J.P.

1970   The political impasse in farm support legislation. Pp. 186-198 in Interest Group Politics in America, R.H. Salisbury, ed. New York: Harper and Row.

Hirshleifer, J., and J.G. Riley

1992   The Analytics of Information and Uncertainty. Cambridge: Cambridge University Press.

Huffman, W.E., and R.E. Evenson

1993   Science for Agriculture: A Long-Term Perspective. Ames: Iowa State University Press.

Johnson, G.L.

1986   Research Methodology for Economists: Philosophy and Practice. New York: MacMillan.

Kilman, S.

1998   Crop deregulation is put to test in new rural crisis. Wall Street Journal (November 9):1.

Kirkendall, R.S.

1982   Social Scientists and Farm Politics in the Age of Roosevelt. Ames: Iowa State University Press.

Koffsky, N.M.
    1966   Agricultural economics in the USDA: An inside view. Journal of Farm Economics
            48(2):413-421.
Kunze, J.
    1991   The profession and the public: Agricultural economics and public service, 1920s and 1930s.
            Economics and Public Service: Proceedings of the 30th Anniversary ERS Conference.
            ERS, USDA (AGES 9138) (July):76-83.
Lee, J.E., Jr.
    1993a   Informational Memorandum for ERS Division Directors. Economic Research Service, U.S.
            Department of Agriculture (February 9).
    1993b   Informational Memorandum for All ERS Staff. Economic Research Service, U.S. Depart-
            ment of Agriculture (February 10).
    1993c   Informational Memorandum to ERS employees. Economic Research Service, U.S. Depart-
            ment of Agriculture (February 24).
    1993d   Informational Memorandum for all ERS employees. Economic Research Service, U.S.
            Department of Agriculture (March 1).
Leontief, W.W.
    1971   Theoretical assumption and non-observed facts. American Economic Review 61(5):1-7.
Lesher, W.
    1991   Agricultural policy in the 1980s. Economics and Public Service: Proceedings of the 30th
            Anniversary ERS Conference. ERS, USDA (AGES 9138) (July):109-113.
Lowett, R., ed.
    1980   Journal of a Tame Bureaucrat: Nils A. Olsen and the BAE, 1925-1935. Ames: Iowa State
            University Press.
Lowi, T.J.
    1962   How farmers get what they want. Pp. 132-139 in Legislative Politics USA. Boston: Little
            and Brown.
Madigan, E.G.
    1992   Statement of the Secretary of Agriculture Before the House Committee on Agriculture.
            Office of Public Affairs, USDA, (June):1, 2.
Mathews, J.T.
    1997   Power shift. Foreign Affairs 76(1):50-66.
McAfee, R.P., and J. McMillan
    1996   Analyzing the airwaves auction. Journal of Economic Perspectives 10(1):159-176.
National Research Council
    1981   Rural America In Passage: Statistics For Policy. Dorothy M. Gilford, Glenn L. Nelson,
            and Linda Ingram, eds. Committee on National Statistics. Washington, D.C.: National
            Academy Press.
National Science Foundation/Science Resource Studies
    1996   Survey of Federal Funds for Research and Development.
    1997   Survey of Federal Funds for Research and Development.
    1998   Survey of Federal Funds for Research and Development.
National Science Foundation
    1998a   National Patterns of R&D Resources 1997 Data Update (http://www.nsf.gov/sbe/srs/
            natpat97/start.htm).
    1998b   Master Government List of Federally-Funded Research and Development Centers (http://
            www.nsf.gov/sbe/anno96/start.htm#mas).
Office of Management and Budget
    1998a   Budget of the United States Government: Fiscal Year 1999. Budget Volume, pp. 261-291,
            361, 366.

1998b   Budget of the United States Government: Fiscal Year 1999. Historical Tables Volume, pp.275-278.

1998c   Budget of the United States Government: Fiscal Year 1999. Analytical Perspectives Volume, pp. 402-403.

Office of the Secretary

1994   Reorganization of the U.S. Department of Agriculture (Secretary's Memorandum 1010-1). (October 20). Washington, D.C.: U.S. Department of Agriculture.

Porter, R.

1990   The Competitive Advantage of Nations. New York: Macmillan.

Potter, C.

1998   Against the Grain: Agri-Environmental Reform in the United States and the European Union. New York: CAB International.

Rasmussen, W.D., and G.L. Baker

1972   The Department of Agriculture. New York: Praeger.

Rasmussen, W.D.

1991   The Economic Research Service: Thirty years of research and service. Economics and Public Service: Proceedings of the 30th Anniversary ERS Conference. ERS, USDA (AGES 1938) (July):84-91.

Reichelderfer, K.

1985   Do USDA Farm Program Participants Contribute to Soil Erosion? (AER 532). Economic Research Service. Washington, D.C.: U.S. Department of Agriculture.

Rodrik, D.

1997   Sense and nonsense in the globalization debate. Foreign Policy 107 (Summer):19-37.

Ruttan, V.W.

1984   Social science knowledge and institutional change. American Journal of Agricultural Economics 66(5):549-559.

Schuh, G.E.

1974   The exchange rate and U.S. agriculture. American Journal of Agricultural Economics 56(1):1-13.

1991   Open economies: Implications for global agriculture. American Journal of Agricultural Economics 73(5):1322-1329.

Schultz, T.W.

1954   Some guiding principles in organizing agricultural economics research. Journal of Farm Economics 36(1):18-21.

Seaborg, D.

1991   Situation and Outlook. Economics and Public Service: Proceedings of the 30th Anniversary ERS Conference. ERS, USDA (AGES 9138) (July):25-29.

Service, R.F.

1998   Chemical industry rushes toward greener pastures. Science 282(5389):608-610.

Stam, J.M., S.R. Koenig, S.E. Bentley, and H.F. Gale, Jr.

1991   Farm Financial Stress, Farm Exits, and Public Sector Assistance to the Farm Sector in the 1980s. (AER 645) Economic Research Service. (April):4-23. Washington, D.C.: U.S. Department of Agriculture.

Taylor, H.C., and A.D. Taylor

1952   The Story of Agricultural Economics in the United States, 1840-1932. Ames: Iowa State College Press.

Taylor, H.C.

1992   A Farm Economist in Washington, 1919-1925. Madison, WI.: Department of Agricultural Economics.

Tenny, L.S.

1947   The Bureau of Agricultural Economics: The early years. Journal of Farm Economics 29(4):1017-1026.

Townsend, T.
   1987   This high-powered instrument that sent its projectile to the vitals of the industry: Why the
           USDA does not forecast cotton prices. Pp. 374-377 in Proceedings of the Beltwide Cotton
           Production Research Conferences. Memphis, TN: National Cotton Council of America.
Truman, D.B.
   1965   Presidential executives or congressional executives. Pp. 404-410 in The Governmental
           Process. New York: Knopf.
Tweeten, L.
   1979   Foundations of Farm Policy. Lincoln: University of Nebraska Press.
   1989   Farm Policy Analysis. Boulder, CO: Westview.
   1992   Agricultural Trade. Boulder CO: Westview.
U.S. Department of Agriculture
   1997   USDA Strategic Plan 1997-2002 (http://www.usda.gov/ocfo/strat/index.htm).
U.S. Office of Personnel Management
   1996   Economist Series GS-0110, HRCD-1, Human Resources Systems Service, Office of Clas-
           sification. April 1996.
Wells, O.V., J.D. Black, P.H. Appleby, H.C. Taylor, H.R. Tolley, R.J. Penn, and T.W. Schultz.
   1954   The fragmentation of the BAE. Journal of Farm Economics 36(1):1-21.
Wilson, J.
   1912   Annual Report of the Secretary of Agriculture. Washington, D.C.: U.S. Department of
           Agriculture.
Womack, A.W.
   1991   National and International Dimensions of Situation and Outlook Work Over Time. Eco-
           nomics and Public Service: Proceedings of the 30th Anniversary ERS Conference. ERS,
           USDA (AEGS 9138) (July):32-35.

Biographical Sketches

JOHN F. GEWEKE (*Chair*) is professor in the Department of Economics at the University of Minnesota and adviser to the Federal Reserve Bank of Minneapolis. He was previously director of the Institute of Statistics and Decision Sciences at Duke University and professor in the Department of Economics at the University of Wisconsin. He is currently a member of the National Research Council's (NRC) Commission on Behavioral and Social Sciences and Education and is a former member of the NRC's Committee on National Statistics and the Committee on Population's Panel on the Demographic and Economic Impacts of Immigration. He is a fellow of the Econometric Society and the American Statistical Association. His research has included time series and Bayesian econometric methods, with applications in macroeconomics and labor economics. He has a B.S. from Michigan State University and a Ph.D. in economics from the University of Minnesota.

DENNIS AIGNER is the former dean of the Graduate School of Management and professor of management and economics at the University of California, Irvine. He is also the founding editor of the *Journal of Econometrics* and is experienced in research administration. He served on the National Research Council's Committee on the National Energy Modeling System, which reviewed the U.S. Department of Energy's modeling and forecasting systems. He has a Ph.D. in agricultural economics from the University of California, Berkeley.

JAMES T. BONNEN is professor emeritus of Agricultural Economics at Michigan State University. He served as chair of the National Research Council's (NRC) Panel on Statistics for Rural Development Policy (1979-1980) and on three other NRC panels since 1989. Bonnen was director of the President's Reorganization Project for the Federal Statistical System (1978-1980), president of the American Agricultural Economics Association (1975), and senior staff economist with the President's Council of Economic Advisers (1963-1965). He is a fellow of the American Agricultural Economics Association, the American Statistical Association, and the American Association for the Advancement of Science. He has a Ph.D. from Harvard University.

IVERY D. CLIFTON is the associate dean of the College of Agricultural Economics and Environmental Science and professor of agricultural and applied economics at the University of Georgia. He currently serves on the advisory board of the Trust for Public Lands and is a past member of the board of the American Association of Agricultural Economics. He also worked as an agricultural economist at the Economic Research Service in the early 1970s and served as director of the American Agricultural Economic Association. He has a Ph.D. in agricultural economics from the University of Illinois.

JOSHUA S. DICK (*Senior Project Assistant*) is a staff member of the Committee on National Statistics. His project assignments have included the Panel on Integrated Environmental and Economic Accounting, the Panel to Review the Statistical Procedures for the Decennial Census, and a study on statistical issues in developing cost-of-living indexes for federal programs. He has a B.A. in political science with honors from Florida Atlantic University and served as an intern for U.S. Senator Connie Mack. He is a member of Pi Sigma Alpha, the national political science honor society, and is an Eagle Scout with the Boy Scouts of America.

KAREN HUIE (*Research Assistant*) served as a project assistant for other Committee on National Statistics studies, including one on the Bureau of Transportation Statistics. She has a B.A. from Wellesley College and an M.A. in German and European studies from Georgetown University. She is currently completing an M.L.S. at the School of Communication, Information, and Library Science at Rutgers University.

GEORGE G. JUDGE is professor in the Graduate School at the University of California, Berkeley. He is an econometrician and his research is concerned with developing improved methods of estimation and inference. He was previously on the faculty in the Department of Economics and Agricultural Economics at the University of California, Berkeley. From 1959 to 1986, he was professor in the Department of Economics at the University of Illinois. He is a fellow of the Econometric Society, the Journal of Econometrics, and the American Agricultural Economics Association. He has a Ph.D. in economics and statistics from Iowa State University.

JEFFREY J. KOSHEL (*Study Director*) served as study director for the first year of the project. He currently is the director of state and local initiatives in the Office of the Secretary at the U.S. Department of Health and Human Services. His prior experience includes serving as a senior fellow at the National Governors' Association, as chief of cost estimation for human resource legislation at the U.S. Congressional Budget Office, and as director of Social Services Research at the Urban Institute.

ROBERT C. MARSHALL is a professor and head of the Economics Department at Pennsylvania State University. Prior to this position, he was associate professor of economics at Duke University. He served as a member of the Committee on National Statistics' Panel on Statistical Methods for Testing and Evaluating Defense Systems. His research, using theoretical, empirical, and numerical methods of analysis, has included a broad range of topics such as housing, labor, the expected utility paradigm, and measurements of mobility. His primary research focus is on auctions and procurements with an emphasis on collusion by bidders. He has a Ph.D. in economics from the University of California, San Diego.

CHARLES RIEMENSCHNEIDER is an agricultural economist who is currently the director of the Liaison Office for North America for the Food and Agriculture Organization of the United Nations. He is the former staff director of the U.S. Senate Agriculture, Nutrition, and Forestry Committee and former vice president of Chemical Bank. He has a Ph.D. in agricultural economics from Michigan State University.

ROBERT L. THOMPSON is a sector strategy and policy advisor for agricultural and rural development at the World Bank and a senior adviser at the Center for Strategic and International Studies. An agricultural economist, he has served as president and chief executive officer of the Winrock International Institute for Agricultural Development (1993-1998), as dean of agriculture (1987-1993) and professor of agricultural economics (1974-1993) at Purdue University, as assistant secretary for economics at the U.S. Department of Agriculture (1985-1987), and as senior staff economist for food and agriculture at the president's Council of Economic Advisers (1983-1985). His major areas of work have been international trade and agricultural policy. He is a Fellow of the American Agricultural Economics Association and of the American Association for the Advancement of Science and a Foreign Member of the Royal Swedish Academy of Agriculture and Forestry and of the Ukrainian Academy of Agricultural Sciences. He has a Ph.D. in agricultural economics from Purdue University.

SARAHELEN THOMPSON is professor of agricultural and food marketing at the University of Illinois. From 1992 to 1997, she was interim assistant director of the Illinois Agricultural Experiment Station. She is currently chair of the Food and Agricultural Marketing Consortium and a coeditor of the *Review of Agricultural Economics*. She conducts research on the performance of commodity, transportation, processing, and food markets; the economic role of futures markets; marketing strategies for agricultural and food products; agricultural economic history; and the economic impacts of agricultural information systems. She has a Ph.D. from the Food Research Institute, Stanford University.

ANDREW A. WHITE (*Study Director*) is acting director of the Committee on National Statistics. He is a former survey designer, research staff chief, and executive staff member of the National Center for Health Statistics and was a consulting statistician with the Michigan Department of Public Health. He directed interdisciplinary research in statistical mapping, survey design, and work in customer satisfaction. In addition to his acting director duties for the Committee on National Statistics, he has served as deputy director for the committee and as study director for the Panel on Alternative Census Methodologies and the Panel to Review the Statistical Procedures for the Decennial Census. He has a B.A. in political science and M.P.H. and Ph.D. degrees in biostatistics from the University of Michigan.

Index

A

Academic research, 5, 27, 28, 130, 145
   Division of Statistics and, 33
   Food and Agriculture Policy Research
      Institute, 108-109, 117-118, 119
   historical perspectives, 33, 34-35, 63, 64,
      117-118, 132
   land grant colleges, 34, 38, 64, 65, 132
      Extension Service, 10, 38, 47, 132, 143
   publications of individual ERS researchers,
      104, 112, 113, 123
Accounting and accountability, general, 4, 109,
   110
   see also Evaluation issues
Administration and administrators, 5-9, 10, 14,
      16, 46, 56, 73-77, 87-88, 105, 106,
      116-146
   evaluation, general, 2, 3-5, 8, 91-115
   evaluation instruments, 110
   evaluation of staff, 110-114
   farm management, 33, 34, 36, 40, 61
   principles of, 125-132
   task analysis and management, 56, 108,
      111-112, 113, 114, 123
   see also Organizational factors;
      Procurement
Agency for International Development, 50-51,
      62, 65, 109
Agricultural Act, 71
Agricultural Adjustment Act, 43

Agricultural Adjustment Administration, 38,
   41, 43
Agricultural Marketing Service, 40
Agricultural Outlook, 41, 95
Agricultural production, *see* Production
   indicators and controls
Agricultural Research Service, 6, 10, 21, 40,
   131, 132, 143
Agricultural Resource Management Study, 140
*Agricultural Resources and Environmental
   Indicators*, 99
Agricultural Stabilization and Conservation
   Service, 38
Agriculture and Food Act, 60-61
American Agricultural Economics Association,
   83
American Economic Association, 35
American Farm Bureau Federation, 40
Animal health, 33
Antitrust, *see* Monopolies
Assistant Secretary for Economics, 11, 145
Auctions, 22-23, 26, 27-29, 112, 132
Australian Bureau of Agricultural and Resource
   Economics, 118-119

B

Biotechnology, 20, 77-78, 80
   intellectual property, 22
Budget, *see* Funding

P

Patent Office, 22

Peer review, 7, 28, 111, 113, 123, 136, 137

Personnel, *see* Staffing and staff analysis

Political factors, 3, 7, 8, 9, 12, 72, 73, 82-83, 84, 86-87, 119, 136
  GATT/WTO, 26
  government intervention in markets, 20
  historical perspectives, 32, 35-47 (passim), 51-53, 55, 60-64
  long-term research, 81
  timeliness of research, 1-2, 4, 5, 7, 9, 16, 30, 106, 107, 109, 121-122, 123, 126, 133-134

Poverty, 101
  historical perspectives, 34, 35, 47, 49
  see also Developing countries; Food assistance programs; Food stamps; Welfare

Prices and price supports, 8, 18, 27, 71, 78, 79, 101, 102, 103, 119, 135, 141
  auctions, 22-23, 26, 27-29, 112, 132
  commodity price projections, 108-109
  cooperatives, 37
  electric power, 25, 26, 27-29
  equity issues, 21
  ERS services, 4
  historical perspectives, 33, 35, 36, 37, 40, 41, 43, 47, 51, 52-53, 55, 56, 58, 59, 132
  monopolies, 18, 25, 26, 27-28
  see also Auctions; Inflation; Subsidies

Private sector, 4, 5, 15, 18, 25-26, 68, 79, 83, 85, 118, 119, 120, 128-129, 145
  evaluation process, 109, 114
  government services overlap, 93-94
  secondary data, 8, 83, 94, 119, 134-135, 140
  see also Extramural research; Procurement; Property rights

Procurement, general, 4, 5-6, 7, 8, 123, 126-128, 130-132, 143, 144
  Competition in Contracting Act, 5, 125
  sole-source awards, 5-6, 125, 126, 127, 128, 132

Production indicators and controls, 8, 71-72, 79, 80, 119, 132-133
  historical perspectives, 41, 43, 45, 53, 63, 71-72, 117
  see also Crop data and estimates

Property rights, 1, 13, 20, 24, 25, 79
  foreign ownership of U.S. farmland, 53, 61
  intellectual property, 19, 20

Publications, 41, 73, 86, 95, 99, 104
  Internet, 21, 128
  staff of ERS, 104, 112, 113, 123
  see also Secondary data preparation and analysis; *Situation and Outlook*

Public goods, 19, 22, 28, 94

R

Reagan administration, 56, 63, 65

Recession and depression, *see* Economic cycles

Regional factors, 47, 49, 55, 119, 133
  substate, 14

Relevance of research, 2, 4, 5, 7, 9, 10, 11, 12, 16, 27, 29, 30, 83, 106, 107, 109, 120, 122-123, 124, 130, 131, 133, 134, 136, 137, 140, 143, 145, 146
  historical perspectives, 51, 62, 63

Roosevelt administration, 37-38, 40, 42-43

Rural development, 21, 23(n.5), 80, 81, 90, 100-102, 105
  electric power, 38, 102
  historical perspectives, 37, 45, 47, 49, 50, 52, 61, 117

Rural Electrification Administration/Rural Utilities Service, 38

Rural Housing and Community Development Service, 38

S

Secondary data preparation and analysis, 2, 8, 81, 90, 95, 134-136, 139-141, 142
  costs of, 128-129
  evaluation issues, 94, 95, 107, 134-136
  foreign trade, 41, 95
  historical perspectives, 41, 46, 51, 55, 68
  mission statements, 107
  private sector, 8, 83, 94, 119, 134-135, 140
  public good, information as, 20, 94
  see also Situation and Outlook

Situation and Outlook, 2, 23, 55, 61, 77, 81, 95, 117, 121, 128, 134, 140
  historical perspectives, 36, 46, 55, 61-62

Soil Conservation Service/Natural Resource Conservation Service, 37, 50-51, 53, 65, 66

Sole-source contract awards, 5-6, 125, 126, 127, 128, 132

Soviet Union, 51